Innovation and Environmental Risk

Edited by Lewis Roberts and Albert Weale
University of East Anglia

Belhaven Press
London and New York

First published in Great Britain in 1991 by
Belhaven Press (a division of Pinter Publishers),
25 Floral Street, London WC2E 9DS

For enquiries in North America please contact
PO Box 197, Irvington, NY 10533

British Library Cataloguing in Publication Data
A CIP catalogue record for this book is available from the British Library

ISBN 1 85293 156 6

Library of Congress Cataloging in Publication Data
A CIP catalog record for this book is available from the Library of Congress

Typeset by Witwell Ltd, Southport
Printed and bound by Biddles Ltd

Innovation and Environmental Risk

Contents

Authors and Contributors

Dr Peter Allen is Joint Head of the Industrial and Environmental Psychology Unit of the Robens Institute in the University of Surrey, and an honorary fellow of the Centre for Health Service Studies at the University of Kent. His research interests include social perceptions, the understanding of risk, and social decision making.

Dr D. J. Ball is a Senior Research Associate in the School of Environmental Sciences, University of East Anglia. He was previously the Head of Materials Sciences Division, London Scientific Services and Head of the Pollution Monitoring Group of the Greater London Council Scientific Branch. He is a member of the Council of the National Society for Clean Air and a Fellow of the Institute of Physics and of the Royal Society of Health.

Prof. J.E. Beringer is the Head of the Department of Botany at the University of Bristol. He is a member of the Health and Safety Executive's Advisory Committee on Genetic Manipulation and the Chairman of the Department of the Environment Health and Safety Executive Advisory Committee on Releases to the Environment. He serves on national and international committees concerned with many aspects of biotechnology.

Prof. Martin Bobrow is Prince Philip Professor of Paediatric Research at the United Medical and Dental Schools, Guy's Hospital, University of London, and Director of the South East Thames Regional Genetics Centre at Guy's Hospital. He worked in the MRC Population Genetics Unit, and the University Genetics Laboratory, in Oxford, and was then Professor of Human Genetics in the University of Amsterdam, before moving to Guy's Hospital. His interests include human and clinical cytogenetics, and molecular genetics. He is Chairman of COMARE, the Committee on Medical Aspects of Radiation in the Environment; and of ULTRA, the Unrelated Live Transplant Regulatory Authority.

Sylvia Brown is a Lecturer in the Open Business School in the Open University and is an Applied Psychologist with an interest in systemic aspects of cognitive processes. Recent research includes the level and status of environmental considerations as inputs to organisational strategy.

Prof. Derek Burke is the Vice-Chancellor of the University of East Anglia. He was previously Professor of Biological Sciences in the University of Warwick and Scientific Director of Allelix Inc., Toronto. He is currently the President of the Society for General Microbiology and the Chairman of the Advisory Committee on Novel Foods and Processes.

Dr Susan Carr is a Lecturer in the Systems Group in the Open University. A graduate in agricultural science, with a doctorate in conservation practice, she specialises in the behaviour and attitudes of farmers to wildlife conservation.

Dr Roger H. Clarke is the Director of the National Radiological Protection Board. He is a member of the Main Commission of the International Commission on Radiological Protection and the UK Alternate Representative to the UN Scientific Committee on the Effects of Atomic Radiation. He is Chairman of the OECD Nuclear Energy Agency's Committee on Radiation Protection and Public Health, a member of the Article 31 Group of Experts under the Euratom Treaty, and has acted as a consultant to the World Health Organisation and International Atomic Energy Agency. He has published widely in scientific literature and at conferences, and produced a monograph entitled *Carcinogenesis and Radiation Risk: A Biomathematical Reconnaissance*.

J. C. Felton has worked for the past twelve years in the Environmental Affairs Division of Shell's central Health, Safety and Environment Division where he is concerned with the environmental acceptability of the whole range of Shell products. He has been engaged in all stages of the evaluation of the development of insecticides and in investigations of agricultural and public health problems in various parts of the world.

Prof. D. L. Georgala, CBE, is the Director of Food Research of the Agricultural and Food Research Council and Director of the Council's Institute of Food Research, the main public institute concerned with food quality and safety and new technology. He was previously Head of the Unilever Research Laboratory at Colworth House.

Dr. R. V. Kemp is a Lecturer in Environmental Planning and in Environmental Risk Assessment in the School of Environmental Sciences at the University of East Anglia. He took part in a review of the Sizewell B Inquiry and directed an independent analysis of the discussion document on the deep disposal of radioactive waste circulated by UK Nirex Ltd in 1987.

Geoffrey Lean is environment correspondent on the staff of *The Observer*.

Prof. T. O'Riordan is a Professor in the School of Environmental Sciences at the University of East Anglia. He is a member of the Norfolk Broads Authority and an author and broadcaster on environmental subjects.

Baroness Platt, CBE, F. Eng., is widely known for her work in higher education and the technical education of women. Trained as an aeronautical engineer, she was formerly a member of the Engineering Council and is currently a member of the Engineering Training Authority and a Board Member of British Gas. She is a Freeman of the City of London and was created a Life Peer in 1981.

J.D. Rimington, CB, is the Director General of the Health and Safety Executive, on loan from the Civil Service where he holds the rank of Deputy Secretary in the Department of Employment Group. He was previously attached to the Board of Trade, the Treasury, the Diplomatic Service, the Department of Employment, the Department of Trade and Industry, the EEC and the Manpower Services Commission.

Prof. L.E.J. Roberts, CBE, FRS, is the Wolfson Professor of Environmental Risk Assessment at the University of East Anglia, and Head of the Environmental Risk Assessment Unit in the School of Environmental Sciences. He was formerly the Director of the Harwell Laboratory of the UK Atomic Energy Authority.

Dr J. Sheail is a Senior Principal Scientific Officer at the Monks Wood Experimental Station of the Institute of Terrestrial Ecology (Natural Environmental Research Council). An historical geographer, he is the author of books on the history of the wild rabbit, the nature conservation movement, rural conservation in inter-war Britain, the impact of pesticides on wildlife, and the development of British ecology.

Dr Joyce Tait is a Senior Lecturer in the Systems Group and Director of the Centre for Technology Strategy in the Open University. She is also a senior member of Wolfson College, Cambridge. She is a graduate in Pharmaceutical Chemistry and holds a PhD in Land Economy from the University of Cambridge. She was seconded, part-time, for three years to the Economic and Social Research Council to co-ordinate a major research programme on environmental issues.

Prof. Dr G. R. Wagner is Professor of Business Economics at the University of Essen, Germany. He holds the first German university chair on the relationship between the business firm and the environment, and is the author or editor of a number of works on environmental economics, most recently *Unternehmung und ökologische Umwelt.*

Prof. A.P. Weale is Professor of Politics at the University of East Anglia. He is the author of publications in the fields of health and environmental policy.

Prof. R. Williams is Professor of Government and Dean of the Faculty of Economic and Social Studies at the University of Manchester. He has written extensively on science policy and has served as an adviser to the House of Lords Select Committee on Science and Technology.

Prof. Harold Woolhouse, formerly Director of the John Innes Research Institute, Norwich, England, is Director of the Waite Agricultural Research Institute of the University of Adelaide. When Roseworthy Agricultural College merges with the University of Adelaide in 1991 he will also be Dean of the new Faculty of Agricultural and Natural Resource Sciences. Professor Woolhouse's scientific background is mainly in plant physiology, biochemistry and genetic engineering.

Preface

'What has been is what will be, and what has been done is what will be done; and there is nothing new under the sun.' These words may have been true when they were written some 2,300 years ago, but they no longer describe the world in which we live. Since the beginnings of the industrial revolution in the latter part of the eighteenth century, the use of science and technology to produce innovative products, materials, foods, chemicals, sources of energy, modes of transport, and manufacturing processes has proceeded with increasing pace and momentum. Even those who are no more than middle aged would no longer feel at home in the world in which they were born, without extensive plastics, preserving agents, domestic appliances, ubiquitous air transport, and communications and information technology. Behind these visible symbols of innovation there lie the less visible, but potent, changes in production technology that have done so much to transform our world, and of which new developments in biotechnology and genetic engineering are simply the latest manifestation.

The great Austrian economist, Joseph Schumpeter, described capitalism as a process of 'creative destruction'. What he meant by this was that the fundamental impulse that sets and keeps the capitalist engine in motion comes from such things as new consumer goods, new methods of production, new markets and new forms of industrial organisation that are the essence of capitalism. In other words the fact of innovation is not something that is incidental or contingently related to the economy in which we live, but is instead part of its defining principle. Without this constant process of innovation, consumers would not be able to enjoy the high standard of living that the citizen of the modern liberal capitalist state has come to expect. The revolution of rising expectations is not under way; it has already taken place.

Yet this central fact of the modern economy, from which the whole of contemporary society takes its colouring, contains its own ambiguities. As basic requirements are increasingly met, and horizons widened, the citizens of the modern economy have also come to expect greater safety in product design and use, less risk to the environment and human populations, a greater regard for other species, and an improvement in the 'quality of life' as well as an increase in consumption. Innovation threatens these values. As case studies of pesticides, biotechnoloy and radiation hazards show, it is all too easy for innovations that promise great benefits to become sources of danger and anxiety both to the lay public and to the experts who are familiar with their principal features. Moreover, with increasingly high levels of education in the population, there is more scepticism and less willingness to trust experts or those in authority so that the dangers of innovation require proper assessment and appraisal before they are incurred.

So this is the situation. On the one hand, innovation promises and delivers great benefits. On the other hand, it is increasingly perceived to threaten the values of health, safety and environmental protection that are so central to the outlook of citizens in the affluent countries of the world. It was with this dilemma in mind that the Interdisciplinary Committee, the Environmental Risk Assessment Unit and the Centre for Public Choice Studies at the University of East Anglia decided to hold a conference to explore the benefits and the risks of innovation. The present volume contains the papers given at that conference, revised in the light of discussion and subsequent reflection. After a session devoted to general issues, several case studies were examined in greater detail and the seminar ended with a general discussion of the main issues that had emerged. The reader's attention is drawn to four themes that appear to pervade the contributions and the discussion.

The first of these themes is that successful innovation requires an *improved quality of understanding* of its effects and implications. The need for this improved understanding is of especial importance among those responsible for taking policy decisions on innovation, but it is also important for the public at large. In this respect the media have a crucial role to play. The conditions under which news is made mean that distortion often attends the reporting of events. More subtly, the commercial imperatives of the media mean that although they can deliver instant information, they are less able to place issues of health and safety in a proper perspective and public perception takes place in a vacuum of understanding.

Issues of the quality of understanding are not simply a matter for the public, however; they are also crucial for policy makers. Because of the complexity and highly specialist nature of the decisions that have to be made, expertise is a precondition for rational decision making in an increasingly wide range of policy fields. However, if lack of understanding is the temptation of the politician and civil servant, an exaggerated sense of the scope and value of specialist expertise is the temptation of the adviser. In modern states the role of the scientific adviser requires a mixture of technical knowledge and socio-political understanding, as well as humility in the face of the complex decisions that have to be made.

No matter how we improve our understanding, however, there will always be uncertainty in the face of innovation. This is our second major theme: *uncertainty is a perennial condition in the face of innovation.* Such uncertainty is not always small or residual; it can be large and pervasive. There is no way in which all the effects of innovation can be reliably predicted and there are a number of consequences that can be expected to flow from this fact. Firstly, there is always the danger of selective perception in the face of pervasive uncertainty. Secondly, uncertainty means that improvements in the quality of understanding may emerge from a better analysis of known facts than from new facts themselves. Thus, much more effort needs to be devoted to analytic issues, if the regulation of the risks of innovation is to be conducted on a sound basis.

Our third theme follows from the previous two. Given our shortcomings in understanding and the perennial condition of uncertainty, *there will always be a need for the exercise of responsible judgement on the part of those who make corporate and public policy.* In such a situation, innovators themselves must internalise a sense of responsibility for the environment itself without the prompting of public regulation if long-term prosperity is to be safeguarded.

Just as firms cannot always pursue short-term efficiency, so the public regulatory agencies cannot always pursue technical effectiveness alone in their regulatory strategies. In addition, they must always be responsive to the concerns of the public. No public agency in a democratic society can afford to ignore public anxiety. Responsibility in such a situation involves acknowledging and acting upon the concerns of the public, even when to the professional these concerns may seem irrational or misguided.

The fourth theme brings together the other three. For it is only by improving understanding, acknowledging uncertainties and behaving responsibly in the face of competing demands, that policy makers and the institutions of which they are a part can secure public legitimacy for their decisions. *Innovation requires a structure of public legitimacy if its benefits and its hazards are to be adequately balanced against one another.* Legitimacy in turn rests upon openness and accountability in decision making and upon a framework of trust between the public and those responsible for making decisions. Where these conditions are absent, there can easily emerge a situation in which confrontation and uncreative tension dominate decision making.

As with all discussions of major public issues, this volume will probably raise more problems than it will offer solutions. But that is surely appropriate. For if any message comes through clearly from all the contributors, it is that the public acceptance of innovation is not a problem that can be dealt with hastily or unreflectively, but is instead an issue that deserves to be at the forefront of public discussion and debate for years to come. With so much new under the sun, it would be foolish to expect otherwise.

Acknowledgements

We have many people to thank for their help in connection with this volume. Our first debt is to our sponsors for responding so positively to the not-so-innovative idea that they might be willing to share the costs of the conference on which this set of proceedings is based. The companies involved were: AEA Technology, BP, British Coal, National Power and Unilever. The NERC and our two Schools contributed substantial sums as underwriters. Without all this support the venture could never have gone ahead.

The original idea had been mooted by the Interdisciplinary Committee, and we are grateful to Dick Crampton, John Allen and John Bowman, as members of the Committee, for all the work that they put into the organisation of the conference. We were delighted by the response of the many distinguished people who were invited to act as chair of the sessions or as authors of papers for the time and effort they devoted to this seminar. We should like to thank particularly Lady Platt for taking the time out of her busy schedule to attend and deliver the keynote address, and to the Vice Chancellor of UEA, Prof. Derek Burke, for opening the seminar and contributing an after-dinner speech which is reprinted in this volume.

Alison Clarke (now Hann) efficiently and effectively took responsibility for the day-to-day organisation involved in the run-up to the conference. David Ball compiled an efficient and sensitive account of the final discussion. We owe a great debt to Carol Forward, Anne Martin and Christine Flack for preparing so much of the typescript at short notice, and to Allison Lewis for computing assistance when converting individual contributions into a common wordprocessing format threatened to become impossible. Finally, Iain Stevenson combined commitment and efficiency in the tasks of publication.

Lewis Roberts
Albert Weale
University of East Anglia, July 1990

CHAPTER 1

The Public Acceptance of Risk and Innovation

Baroness Platt of Writtle

Thank you for inviting me to give the keynote speech at this important conference. I must first of all compliment the University and the Interdisciplinary Committee on arranging a seminar on this vital subject, which brings many faculties' expertise to bear on the problem, which is essential, and secondly organising such an interesting and varied programme of speakers which I look forward to hearing very much.

When I received your invitation my first reaction was 'Why me?' and I felt a sense of inadequacy as an out-of-date woman engineer. Then, as I began to think about it, the fascination of such a challenging subject began to take hold of me. You could describe writing today's speech as an oxymoron – a bitter-sweet experience. Rather like the subject we are tackling, in such a situation one can oscillate wildly between extreme fear and the sense of a great challenge.

In circumstances like these one cannot do better than return to first principles in the dictionary. Innovation is defined as the introduction of novelties, the alteration of what is established, and in 1598 an innovator is described as a revolutionist. Risk is described as a hazard or danger, exposure to mischance and peril, or in 1719 the hazard of commercial loss. Johnson is quoted as saying 'To risk the certainty of little for the chance of much'.

From birth human beings have always led lives subjected to risk. In primitive times risks were great: think of the first person who rubbed two sticks together. The likelihood of survival was low, and few lived to old age. Today medical and engineering advances have led to a tremendous reduction in infant mortality in this country, and an immensely enhanced likelihood of living into old age. Indeed one of the problems of the twenty-first century may very well be how the working population is to support an increasing number of frail elderly people.

We are talking today about technological and scientific innovation, and in many cases the public would welcome that innovation with open arms. The surgical operation to insert artificial hips is now seen to be overwhelmingly beneficial, and there are queues of people waiting to have it done.

Replacement knees are seen to be risky, as were hips only a few years ago. The public perception of hip replacement has changed radically, and they now feel comfortable with it.

During the Second World War I worked as a very young woman engineer just down from Cambridge in the experimental flight test department of Hawker Aircraft. I remember very clearly those first flights of Hawker prototypes. They had been designed at Kingston and built there at the start. Then they came to the experimental flight test department for the skilled workmen to finish them off, essentially as a one-off job although later they were destined for mass-production. They had been drawn by hand, hundreds of detailed layouts, many modifications being made as they were built, and all the calculations made by slide-rule by individuals in the design office. Sydney Camm as Chief Designer believed in designing them pretty and simple and strong, with generous safety factors in stressing. Later in their flying lives the aircraft proved capable of carrying much greater loads with safety than those for which they had been designed, but those first flights were very cautious. The Chief Test Pilot was in the pilot's seat, all the top brass was there, from Tommy Sopwith the Chairman downwards, and we stood out on the tarmac in great excitement in the wind watching the preliminary taxiing flights, and then accelerations across the grass airfield until eventually that magical moment when the aircraft was free of the shackles of earthly existence and she entered into her natural element, the air. A few circuits and bumps and we could then all look forward to months of testing, so that we knew eventually her top speed and how it varied with height, her fuel consumption, rate of climb and many other important factors such as how she behaved in fast dives near to the speed of sound. The sound barrier had not then been broken, and pilots described the sensation as like coming downstairs on a teatray. G-suits had just been invented so they formed part of the testing, at the same time testing out the G-suits' own efficiency, which prevented pilot blackouts on tight turns. An aircraft had crashed killing the pilot just before I arrived because of the effect of negative g on the carburettor. Very soon after that the engines were fitted with injectors to overcome that problem. Throughout that demanding testing programme, designers, test pilots, technicians, and ground staff worked very closely together. We worked very hard, were fiercely proud of our aircraft and determined they should be as safe as possible, fast and successful in battle. We had a war on our hands.

During the latter part of the war I was involved with the experimental flight testing of the single-seat fighter the Tempest V, designed to destroy enemy fighters and bombers with devastating fire power, rockets, bombs and cannon. When it finally went into service it arrived just at the time of the attack on this country of the doodlebugs, which aroused fear in all hearts as their engines cut out and you knew they were about to crash, explode and kill. The Tempest V's were at the leading edge of innovation, very fast, and the pilots at great personal risk developed the skill of flying alongside the doodlebugs over the Channel and tipping their wings. The doodlebugs' gyros were not good enough to regain stability and they fell harmlessly into the water – a fascinating example of how a machine designed to kill in war in fact succeeded in saving thousands of civilian lives. The side-effects of innovation are difficult to prophesy.

In fact this short anecdote gives two ways of looking at the title of our seminar: either as the public acceptance of risk when it leads to innovation, or the acceptance of innovation which may in itself minimise risk.

After the war in the late 1940s I worked in the air transport industry, in British European Airways. At that time we weighed all the passengers as well as the baggage, incredible to believe now. My boss decided we should make a statistical analysis of weights of passengers over a period to decide whether we could cease weighing them on normal scheduled flights, in aircraft above a certain size. We were not so concerned with the overall weight as the movement aft of the centre of gravity of the aircraft if by chance all the heaviest passengers were seated in the after end of the aircraft, which might then go into a spin. After a great deal of work based on slide-rule calcuations we decided we could cease weighing the passengers safely. I remember our final advice to cabin staff was that if an abnormally heavy passenger boarded the aircraft they should seat them over the wings. Today the idea of weighing passengers seems archaic, and the sheer logistics of the operation impossible.

We were very concerned with air safety, and at that time the International Civil Aviation Organisation had just been set up on a provisional basis, so that we had to work towards fulfilling their safety regulations over a period of time. All aircraft in future had to have a rate of climb on one engine after the other had failed, but we had a motley fleet of DH Rapides where that was certainly not the case. I was sent up as the technical assistant to measure the rate of descent over several thousand feet. We naturally carried out this experiment several times so as to produce a more accurate average and the pilot and I were glad each time when the plugs had not oiled up and the engine roared into life again for the next experiment.

In obedience to the ICAO regulations I had to do an exercise covering the whole BEA route operations to check that on engine failure and with reduced rate of climb the aircraft could reach an alternative airport, perhaps over mountains, with generous safety margins, and adequate fuel reserves. In hot and cold temperatures at high altitude airports I also worked out at what speed the aircraft should continue taking off after engine failure. At low speeds I had to check the aircraft had plenty of room to land again within the runway lengths. Tremendously detailed work had to be done including checking all the obstacles around European airports, such as gasometers, high buildings, power lines, etc. As you can see risk was at the forefront of our minds in setting up a new airline. In those days our motto was 'The customer is always right'. As engineers we were determined to serve them well.

Twenty-one years ago this July the first men landed on the moon – 'One small step for a man, a great leap for mankind'. As a child, space travel for me was science fiction. Now it is history. It will always be dangerous but men and women with a spirit of adventure are prepared and even eager to take the risk. This courage leads to great benefits for us all, in new materials, medical advances and improved communications. It must never be forgotten that behind the spectacular take-off high into the unknown with all the razzmatazz for the space travellers, lie months and years of patient work by the design team of engineers. They only reach the headlines if something goes wrong, which fortunately it seldom does. Today they will be working with computers not slide-rules, which allows modelling of a great variety of situations in order to assess risks and also to find optimum solutions to problems. My mind boggles at the complexity of such an operation but they of course will build now on past experience in space shuttle operation and will face up to the juxtaposition of worst situations in different fields of risk, so as to do their best to avoid disaster. Today that will involve immense cost on its own, but that cost fades into insignificance compared with the cost of the crash of the Challenger.

In assessing risk of course we cannot forget its other definition, the financial one. When having to sell in the markets of the world, if the price escalates beyond reason, as a result of unreasonable over-caution, the product will never be made as no one will be able to afford to buy it. Nevertheless the manufacturer will know that a reputation for unsafe products can never be overcome. A chain of responsibility has to extend in all directions if safe products of an innovative nature are to be produced safely. The designer has to explore the risks to the full, test the prototype on the bench for long periods to find the effect of wear, test the prototype in action in whatever medium it operates, and create the most risky scenarios under test conditions to find out the weak points in design long before the article goes into production, so that the necessary modifications can be made to promote safety. After that, organisations like the Health and Safety Executive, whose contribution appears later (Chapter 2), the British Standards Association, and the Royal Society for the Prevention of Accidents, ICAO and many others will have regulations which have to be met, and comments that demand attention.

The engineer and designer have great responsibilities to fulfil in this field, as have the executive directors and board of a firm. Once decisions have been made, the concept of customer care and personal responsibility for safety needs to permeate the whole workforce, and that is a very difficult matter, demanding clear communication systems being set up to travel throughout the firm's hierarchy. Training in the maintenance of good standards of safety will be essential, and indeed in these days of constant developments in innovative technology, that will mean training in the understanding of new technology and its applications at appropriate levels throughout the workforce. Otherwise fear of the technology by the workforce and lack of knowledge of the results of mistakes in workmanship can lead to unacceptable standards of risk for the customer.

Increasingly at all levels we will need well-founded initial education of both the workforce and the customers in the applications and use of innovative technology so that they become far more knowledgeable. The new national curriculum will do that in time, with its insistence on the study of science and design and technology throughout school life to the age of sixteen, with emphasis on practical experiments and project work. All children, girls as well as boys, need to be presented with intractable practical problems to solve to the best of their ability, to prepare them for the problems they will meet in the future during their adult lives.

Too often in the past it has not been thought necessary for girls to have this practical technical knowledge but today it is essential. They may in fact become professional engineers themselves and indeed will have to do so if we are going to have the number of engineers available to spearhead technological advance in the twenty-first century, and keep up with our international competitors and indeed keep ahead of them wherever possible. With the fall in the number of eighteen year olds from 900,000 to 600,000 between 1981 and 1995, far more use must be made of the feminine half of the brains of the nation than has been the case in the past, if skill shortages are to be avoided in technology-based fields of work. That was the whole point of the WISE campaign (Women into Science and Engineering) set up in 1984 by the Engineering Council and the Equal Opportunities Commission. I am pleased to say it has received considerable support from a great variety of people and institutions, and women now form about 12 per cent of first-year undergraduates in engineering, but it must continue for a decade if it is to succeed in the long term. I should like to say much more about the WISE campaign, which, as is well known, is very dear to my heart, and vital to the future prosperity of our country. However I am going to control my enthusiasm in that direction and speak about innovations the Engineering Council and others have set up to awaken a love of and interest in science and technology in all children, boys and girls.

The Association for Science Education, the Standing Conference of Schools Science and Technology, Industry Matters and many others have set up technology projects to inspire young people to take up the challenge of solving problems which inevitably arise when setting up altruistic projects to help other people which children very much enjoy. The example I am going to describe is one of thousands. This arose from a project backed by the Nuffield Foundation and Professors Black and Harrison who set up the Technology Education Project. It is the Peruvian Water Project run at Deyes High School, Sefton, with third year pupils.

The idea was sparked off by some material from Christian Aid, drawing attention to the problems of water shortage in a Peruvian village. The scope was wide, and a number of departments became involved: religious education, science, English, art, geography and CDT (craft, design and technology). All had expertise to offer and could tackle various aspects of the project. The normal timetable was suspended for three days and the pupils, placed in small groups, were presented with the same brief – to suggest ways of solving the water shortage. The eventual outcome was a proposal for a reservoir. The groups then looked at the implications of their decision. Some looked at testing water pressure, others at porosity of soils, water purification, hydroelectric power, etc. Role play was used to help the pupils understand the problems of the village.

In this way, the pupils were led into diverse problem areas, often requiring them to identify (with appropriate guidance) the knowledge and skills necessary to produce solutions. They now found that they not only had to acquire new knowledge and skills, but they had to call on existing knowledge hitherto unutilised; and before they could use this existing theoretical knowledge, moreover, they had to transform it into a practical form. They often did not realise that problems existed until they had been prompted by the staff. Solutions to problems seemed easy until the consequences of their decisions were pointed out. What were the social implications of flooding someone's home to provide water for others? Then pupils had to balance one option against another, in the same way as decisions are made in the real world.

The students recall the project with enthusiasm. They felt they had used real knowledge to solve real problems. In the course of their work they had come across many incidental problems, all of which had to be solved in order to proceed. They gained insights into the difficulties facing poor countries, and had developed skills of communication and research. In a way it is invidious to quote one example but I have not time for many and I felt this one described vividly the broadminded approach we must wish to see encouraged.

Another important initiative of the Engineering Council developed from the 'Opening Windows' scheme whereby engineers, young and old, go into schools to speak to children of the excitement, challenge and satisfaction of their work, to encourage the children themselves to consider a future career in engineering. With the present shortage of engineering skills exacerbated by the steep downward trend in the overall numbers of school leavers, that scheme is over the next few years going to develop into the Neighbourhood Engineers project. When that is fully operational three or four engineers will be attached formally to each school to co-operate with teachers to enable their students to have far more practical experience and knowledge of what an engineering career means, and we hope to awaken their enthusiasm for careers in the profession. I hope very much it will be successful, but it will demand much voluntary effort on the part of engineers, with strong industrial and educational backing.

There will always be purveyors doom and gloom and Luddites who will put forward the argument that they don't understand technology, regard it as an invention of the devil and wish to eschew all contact with it. They will be prepared to suggest that we go back to walking everywhere, or stay at home, and do everything by hand, and stop applying innovative technology. They forget that one of the most dangerous places to be is at home, with more deaths occurring there than on the roads, the next most dangerous place. Far more deaths occur there than at work, and the causes of death and injury at home are most likely to be from simple rather than highly innovative technological causes, such as fire, falling and human error. As the Royal Society for the Prevention of Accidents says, they are almost always preventable, so education in accident prevention during the school days of both boys and girls will be an excellent investment for the future whether at home or at work.

Truth to tell we cannot put back the clock. Danger has always been with us throughout history. As the prayer for a January Sunday says, 'O God, who knowest us to be set in the midst of so many and great dangers, that by reason of the frailty of our natures we cannot always stand upright: grant to us such strength and protection, as may support us in all dangers, and carry us throughout all temptations: through Jesus Christ our Lord. Amen'.

Part of that strength and protection is our duty as a nation to educate all our children to have a proper basis of scientific and technological knowledge on which to found their own code of practice in dealing with hazardous situations at home and at work, and to take proper avoiding and precautionary action to prevent accidents both to themselves and later their own children or elderly relatives. Foresight and precaution, which prevent accidents, are worth more than kind aftercare.

As new innovations with their attendant risks are sold in the shops to the general public, proper instructions for their use need to be made available too. Recently there has been a major failure in communication in the safe use of microwave ovens. In their case it seems familiarity has bred contempt with lack of a proper understanding of possible health hazards resulting from undercooking, which is now being put right. That should have been foreseen and clear warnings given to non-technical users. One has got to accept that mass usage may include people who don't bother to read the instructions, will take short cuts, and expose themselves and their families to unnecessary dangers. Manufacturers cannot be responsible for utter carelessness, but they do need to use their imagination to prevent the worst risks, and of course today there are strict consumer regulations which they have to obey. In a free society we must all respect the law, and consumer protection is now, I am glad to say, given considerable emphasis, forcing manufacturers to consider risk more seriously and avoid it in their innovative products.

As regards safety in the home, both gas and electricity are hazardous products used every day by non-technical people. They have to remember not to touch electrical installations with damp hands, to turn off the current before mending a fuse, to check that they have not turned on the gas without lighting it, and so on and so on. I was lucky in having lessons like that in a girls' school in the 1930s. The suppliers also have responsibilities. When I was a child we used coal gas. Now, of course, it is natural gas, very much cleaner environmentally, non toxic, and with double the calorific value.

British Gas had to take a most risky decision to change over from one to the other in the late 1960s and early 1970s. They had to press government to accept the viability of this immense change and in the event 35 million appliances were changed over in 13 million homes, offices and factories. That was certainly innovation on an immense scale with all sorts of risks, both in delivery and financial ones, involving immense quantities of natural gas from the North Sea. In the event it was achieved relatively smoothly at the rate of about 2 million dwellings a year over a period of less than ten years. Again we look back at it as history, and the use of gas particularly in central heating has gone up by leaps and bounds but at the time there were repeated demands by politicians and the public to stop the conversion programme. Fortunately engineers, government and others in British Gas had satisfied themselves that the risks both financial and otherwise could be overcome, and now the country as a whole benefits from a much better gas supply. The risk in the public eyes is no longer a source of anxiety. That however is due to a great deal of well-informed preplanning, training of conversion operators and foresight so that major difficulties could be overcome.

At the time a high-pressure transmission system also had to be installed to bring the gas in from shore terminals to villages, towns and cities throughout the country. That high-pressure transmission system itself needs testing as millions of cubic metres of gas flow through it daily. British Gas won the MacRobert Award this year for its development of the 'intelligent pig' used in on-line inspection of the pipelines. The pig travels through the pipelines propelled by the gas pressure itself so the pipeline does not have to be taken out of use during testing. As it travels it sends back an immense amount of data so that the pipelines are scanned, and evidence sent back of corrosion or weld failure. When one thinks of the terrible Russian disaster of the leaking gas pipeline near a railway, one can only be grateful for the patient work of gas engineers in the reduction of risk in a highly innovative way using modern technology.

Just at the moment many people in London are suffering from the redevelopment of Liverpool Street Station, another innovation with considerable associated risks. Every day thousands of passengers make their way through what is essentially a hard-hat area. Every few weeks a major feature of the station changes, entrances, exits, passageways to the tube, ticket offices, lavatories – a bewildering labyrinth for the unsuspecting traveller. Nevertheless directions are given, we find our way over rough surfaces and I don't think any passenger has come to serious harm. I would give a medal to the engineer who has worked out the critical path analysis month by month, year by year, and ensured that millions of passengers have travelled to work safely during the period of development, by no means yet finished. Careful thought has obviously been given to the reduction of risk. Let us hope the final innovation is worth it. I wish that British Rail were as innovative in the running of trains as the construction engineers are in preparing the station for the twenty-first century.

During the time I was preparing this speech I asked various people for their personal views, and one of the most vivid I received was from Alan Newton, CBE F.Eng., now retired and a consultant to Rolls Royce. I quote below from what he wrote to me. I am most grateful to Mr Newton for allowing me to quote him and only wish I could include all he wrote, but space does not permit.

Risk and Innovation in Aero-engines

In the aero-engine business, risk may take a number of forms. Today, the customer is able to negotiate a large number of guarantees in the contract such that, if these are not met, then the supplier has to pay. These guarantees may be on fuel consumption, thrust, reliability, component life, maintenance costs, etc. The regulations regarding safety are extremely stringent and are continuously revised and strengthened in the light of service experience. Air transport, by any yardstick, is very safe and is continually improving – even though tragic accidents do occur from time to time. To ensure this standard of safety involves extremely detailed analysis of all components, and testing to confirm that analysis. Service experience is monitored to validate the design assumptions. One tool which is invaluable in modelling the physical processes in an engine is the use of very powerful computers. Since most of the engine is required to work near to the limits of knowledge any innovation is very carefully studied by the designer and the Airworthiness Authority to ensure that it involves no degradation of safety.

Against this background the designer is still faced with the necessity for innovation. There is continuous pressure, both by competition and customer demand, to improve the standard of aero-engines. This may be towards lower fuel consumption, higher thrust, lower weight, better reliability or some combination of these. A major innovation in the business was the advent of the gas turbine engine and this has transformed the aerospace scene:

Concorde flies London to New York in just over three hours compared with fifteen hours for Stratocruisers.
A Quantas 747 has recently flown non stop London to Sydney.
We now have high supersonic military aircraft.
Civil engines now regularly stay on the wing for 20,000 hours without removal, which is a standard of reliability unknown with piston engines.

This is an extreme example of considerable innovation and risk but equally considerable returns.

The high-pressure turbine blade is probably the most difficult design problem in the engine. It operates at gas temperatures higher than the melting point of the blade material and each blade may carry a centrifugal load roughly equal to the weight of a double-decker bus. Even with the improved materials which have been developed over the years it is necessary to cool the blade with (relatively) cooler air tapped from the compressor. This involves a performance loss and so the double demands of higher temperature and better cooling efficiency lead to a need for constant improvement. Again the computer is an essential tool in reconciling the various design aspects – aerodynamic, stressing, vibration and heat transfer – and it is

> significant that we now regularly design blades which achieve more than 20,000 hours in service.
>
> The demand for better engines means that the designer is constantly introducing new features and concepts, i.e. innovation is essential. He or she is conscious of the risks and is using all means to reduce them. It is significant that today's engines are more efficient, more reliable and have a better safety record than their predecessors, which is an indication that he or she is succeeding.

During my presentation so far I have concentrated on everyday innovation planned and carried out by responsible and qualified people whose main aim and object is to carry out a service safely but economically. The problems are experimented with on a considerable scale away from the public, under controlled conditions, so that they are not a danger, and then plans are made to overcome them.

What I hope to illustrate is the sense of responsibility of these people, the innovative turn of mind they display to avoid and overcome risk, and the constraints under which they operate, both of cost and of a regulatory nature, but nevertheless achievement of safety remains paramount for them. I am quoting good examples and others will not be so conscientious and in the end human error is bound sometimes to prevail, but I nevertheless feel it is important to highlight this conscientiousness.

The Engineering Council in its bylaws lays down a simple code of conduct which exemplifies this attitude:

> Every individual on the Register shall at all times so order his conduct as to uphold the dignity and reputation of the engineering profession.
> He shall at all times safeguard the public interest in matters of safety, health and otherwise. He shall at all times exercise his skill and judgement to the best of his ability and discharge his responsibilities with integrity and due technical competence.

The medical profession has its Hippocratic Oath, and other professions have equally strict standards of conduct. Thankfully in innovative situations very few professionals today work on their own. They are part of a multidisciplinary team, and each has to respect the others and know enough of the others' techniques so together they can pool their knowledge and co-operate in minimising risk in the project.

Today some of our public anxieties are on a grander scale – global warming, acid rain, and AIDS and other infections. People feel less in control and are very fearful. Major infections like the plague in earlier times have been uncontrollable, but today we do see the reduction of major diseases like tuberculosis and polio and are always hopeful that medical science will find ways equally to control AIDS and indeed cancer. However we all know that the one thing certain in our lives is that we shall die of something. That will never be avoided. We are mortal and the way of our death is in the hands of God, which we must accept. In my view a little more old-fashioned Christian faith in a loving God would help a great many people, and maybe that will come again. Meanwhile they often sit at home alone and worry about situations they cannot control in a most unconstructive and indeed very demoralising way and get very depressed. I do not think this is helped by the media, who speak often in a superficial way in terms of headlines, and sometimes seem to want to stick to their story against the strict factual evidence which is much more boring to explain. In a free democracy there is no doubt of the need for a free press, radio and television whose duty it is to criticise, and highlight irresponsibility in high places. I do believe nevertheless that they too have a responsibility to give the news in a balanced way so that the public does not become unnecessarily alarmed. They too need to be technically knowledgeable so that they can exercise judgement in presenting the news in a well-informed way. At present there are a number of people in East Anglia living on low-lying ground worried about the flooding of their homes in the near future due to global warming, whereas in fact the danger is many years away, by which time protection could have been provided, as it was along the Thames after the great tide.

Living in a free country is a great privilege, a fact we are made very conscious of when watching the events in Eastern Europe. They are only just enjoying for the first time the sort of freedom of media comment that we take for granted in our country.

During a lifetime in local government I have become a member of many pressure groups, and appreciated their support in various campaigns for better consideration and solutions of local problems. As a Chairman of the Education and later Finance Committee in Essex I was also on a number of occasions at the receiving end of pressure-group criticisms. I nevertheless believe they are an essential part of a democratic country. The price of liberty is certainly eternal vigilance, and the bark of the local watchdogs is of great value, and can result in much better decisions by those in authority. Once again I believe that pressure groups need to act responsibly, amass their information accurately, bring together conflicting views and try to act positively and constructively in co-operation with the statutory body. I realise that means overtures on both sides but I do believe responsible and well-informed stances on both sides result in much better final plans for action. I believe that criticism has to be exercised just as responsibly and knowledgeably as must the people in power themselves control their actions. A very good example of this kind of comment is the present House of Lords Select Committee report on Global Warming, which I think is a shining example of balanced judgement, and should be prescribed reading for everybody anxious to know more about the subject. The Committee recommends action in many areas but at the same time in other areas advises caution. I found it most helpful reading on a subject I cannot pretend to understand in depth myself.

I once again pay tribute to the University in organising this challenging conference and I hope that in my keynote address I have provided enough examples to illustrate my feelings on the subject of risk and innovation. I look forward very much to hearing the speakers and discussion during the rest of the conference which will be devoted I expect to very different points of view. I have spoken as an engineer but all my views of responsibility are generally relevant to all professions. For me risk is inevitable, but those in power must do their very best to minimise it, exercising innovation to find solutions. Innovation itself has to happen but time may very well be a vital ingredient in finding safer ways forward and making the ideas more generally acceptable to the public. The public needs continuing education to keep up to date in a technologically developing world, just as do engineers and the professional managers in industry, commerce, the press, pressure groups and of course government at both national and local level. Nobody when they leave their formal initial education is educated for life. They will need continual updating in technological advances.

Equally people must be educated to act responsibly in a free society in whatever capacity they are operating. Freedom and responsibility are opposite sides of the same coin, and one cannot succeed without the other. That is of course a fundamental truth, especially in the light of a Christian faith. No human society will ever be perfect, so I am putting forward ideals, but perhaps that is what this conference is about. After all each one of us will need a faith by which to live in a difficult world whatever that may be. In conclusion I can do no better than quote from King George VIth's Christmas Broadcast:

> I said to the man who stood at the gate of the year 'Give me a light that I may tread safely into the unknown'. He replied, 'Go out into the darkness and put your hand in the hand of God. That shall be to you better than a light in the darkness and safer than a known way'.

CHAPTER 2

Innovation and the Public Perception of Risk

J. D. Rimington

This book is about public attitudes to innovation, and I think I have been asked to contribute to it because the public is more concerned now than formerly about the hazards attaching to the large-scale exploitation of modern technology. There is now a strong public dislike of industrial consequences that endanger the environment, and a very strong and emotional reaction also – to come nearer to my own sphere – when there are major accidents unexpectedly affecting large numbers of people, some of which, as for example in the case of Chernobyl, can have serious environmental effects also.

I do not think it proven that these strong and natural feelings extend to a hostility to innovation as such. There are certainly people in the industrialised West who harbour some form of 'back to naturism' and a still greater number of people who will for example take photographs or consume tranquillisers while condemning the production and disposal of the refined chemicals that make these activities possible. But, on the whole, the attitude to innovation *per se* seems still quite a healthy one and the resistance where it occurs is to the emergence of technologies or activities that most sensible people would want to ask questions about, even if they did not oppose them.

Since hazard and consequence are what people have in mind when they ask such questions, it is right, at the outset of this Conference, to seek to analyse the state of the public mind (if there is anything so simple) about technological hazard; and I come therefore to the subject of my paper. I intend to suggest some lessons from experience.

The subject is the public perception of risk. To be clear however it is not risk the public is worried about, it is hazard. The difficulty in fact often is that if the hazard is disliked enough, the question 'what is the risk?' receives insufficient consideration; it may indeed be very small. When members of the public are interrogated, by means usually of questionnaires, they often turn out to estimate risks attached to particular hazards differently from experts. One can therefore argue that if a knowledge gap could somehow be closed, the public would behave differently and perhaps in a way more favourable to technological developments which they question.

Part of my purpose is to ask whether this is so; whether, if it is so, the matter is remediable; and whether there are other approaches to the difficulty than filling in knowledge. I must however first explain that I look at the matter from the peculiar point of view of a risk regulator acting for the public, and I had better say what this is.

First, I have to find some way of allocating my resources rationally, and naturally that has to be related to risks. So, for example, out of a total Health and Safety Executive field inspector force of about 1250, over 100 are allocated to construction and over 40 to mining, two very risky industries accounting between them for one-third of the fatalities occurring annually in industry. On the other hand, since we live in a democracy, and my job is to see to the protection of the public as well as workers, I also have to allocate resources according to public concern. The public is particularly concerned about nuclear safety and that can be the only reason why the Executive employs more nuclear inspectors than it devotes to the mining and construction industries combined, since the numbers actually killed by industrial radiation are, on any reckoning we can make, very small.

One might therefore say that risk and the perception of risk are to me broadly what demand and effective demand are to Lord Sainsbury. But the matter goes further than that, and enters into the detail of my business. The principal UK legislation on safety has as its cardinal principle the injunction to do what is reasonably practicable in the interests of safety. Legally what this means is to proceed in the direction of safety until the cost of the next step would be grossly disproportionate to the remaining risk. Generally, what it means is to conform to good practice. Where there is complex nuclear or chemical plant, the production of a numerical risk assessment is implied. Such assessments, which are now carried out for all high-hazard plants, are useful in several practical ways, one being to target subsequent inspections to the riskiest parts; but they have the further effect of leading inexorably to the question, what is your risk standard? How much risk, as measured, will you tolerate? Or would the public tolerate? I shall come back to this in a moment.

Involved in all this is the question of what it is worth paying to avoid very large risks. The investment that has been necessary to reduce the risk of a nuclear accident, and to purify waste streams after reprocessing, has been very great indeed. Though a considerable superstructure of logic has been raised, not least by my own organisation, to justify on a risk-numerical basis the demands that have been made on the nuclear industry in the interests of safety, one could forgive an outsider from wondering whether, in part at least, the demands may have been pressed home more in proportion to the industry's willingness and ability to pay than to any more rational framework of calculations.

The nuclear industry however is not the only area where this question arises; what for example is it worth paying for systems of communication that might render trains immune from accidents such as at Clapham where a failure of the signalling system led to considerable loss of life and injury. And how do we deal with those still more difficult questions where the costs are simply not determinable since they concern not just investment for a determinable return but also reductions in benefit – as for example perhaps the substitution of fossil fuels for nuclear, or for that matter, lead-free petrol? Again, I shall come back to these questions.

Leaving aside these teasers, the erratic nature of the public reaction to hazard has a quite separate significance for a regulator, namely his or her liability to get the blame if something goes wrong. Too much criticism, particularly if it happened to be justified, would reduce confidence in HSE and so its usefulness. Other regulatory bodies have in fact found themselves whether deservedly or not in such difficulties – as the US Nuclear Regulatory Commission did after the Three Mile Island incident, to come no nearer home. It is hardly surprising that public risk regulators should spend some time asking themselves whether and how public perception, or at least reaction, can legitimately be managed, and the same question of course occurs to others whose interests are involved. By way of illustration, it is worth recalling two very different episodes, the businesses of asbestos and of Trawsfynned Power Station.

In the case of asbestos, the dangers of exposure are very real, and were first remarked in the last years of the nineteenth century. The substance possesses great usefulness and is without satisfactory substitutes; quite possibly its fire-resistant properties have saved many more lives than the fibres destroyed. Perhaps these benefits explain why it escaped for so long the strictest forms of regulation world wide. When the subject at last came fully on to the public agenda towards the end of the '70s, the Health and Safety Commission and Executive did not act quickly and firmly enough to allay public concern. There were good reasons, particularly an intention by the European Community to regulate, and there were indeed existing regulations which allowed the heat to be turned on quite considerably. But escalating fears were not dealt with in a manner that caught and satisfied public attention.

In consequence, the action when it came was drastic and the reaction even more so. The stripping of asbestos, much of it wholly unnecessary, must have involved huge expenditures. At the height of the panic, one public man made matters probably worse with the comment that a single particle could kill. The statement is of course logically correct, just as a single straw will break the camel's back; but it has to be considered in the context that it would be difficult to enter any room in the kingdom without risking an encounter with a few particles of this very widespread substance, and that many people have survived without noticeable ill effect very prolonged working exposures to large quantities. The comment and the reaction to it illustrate very well the difficulties of explaining the nature of risks in a popular way; the same problem attaches to radiation.

The episode of Trawsfynned Power Station is of a different order. One of the Sunday papers about three years ago published a photograph which at first sight appeared to be of a skeleton dragging its way across a moonscape, but which on closer view appeared as an engine pulling some wagons across a rickety bridge in the vicinity of some large and gloomy buildings. The text explained that this was Trawsfynned Nuclear Power Station, where in the fastnesses of Wales, the Central Electricity Generating Board was about to conduct an experiment of a dangerous kind inside the reactor vessel.

By the end of the week, a very considerable hoo-hah had set in. The Chief Nuclear Inspector had had to report to me that he had not been able to make himself heard at an invited meeting at the House of Commons because of the local member's constituents, most of whom turned out to belong to an environmental pressure group. Rumour had it – though I have not been able to verify the story – that units of a Lancashire fire brigade actually made arrangements to present themselves at the station to await the outcome of the experiment, in the finest traditions of the Fire Service. If so they were disappointed since the CEGB, in the face of the public outcry, thought it best to cancel the trial.

The Nuclear Installations Inspectorate (NII) whose consent was required had been informed of the proposed trial which was in fact intended to verify whether in the event of six circulator fans not working, the natural circulation of gases within the reactor would have brought about cooling after shut-down without their assistance. The Inspectorate was studying CEGB's proposals at the time the panic took place and would have been present had they permitted the trial to take place. In principle there was no ground for apprehension since the temperature inside the shut-down reactor could easily be monitored and if it failed to reduce as quickly as expected, the fans, the need for which was anyway in doubt, could readily be switched back on. To come quickly to one of the morals of the tale, a main contributing factor in what was really a public relations battle was what the press described as CEGB's secrecy. They had delayed informing their local liaison committee of their proposed test until they had obtained NII's formal approval. The information leaked out, and the pressure groups were able easily to exploit the situation. Both the episodes I have described were failures in the management of public perception – or perhaps from another point of view, famous victories.

These tales remind us of some of the stuff of which the public perception of risk is made up, and particularly of the dramatic elements that are inherent in the subject; and the way that these naturally attract the attention of, and are exploited, by the media. Would it be right though to describe the public's reaction or perception of these risks as in any way irrational? After all, we are conditioned by very basic impulses either to seek to control hazards or to flee from them. The relevant question perhaps is whether a more consistent and reasoned approach might have been obtained by an earlier filling in of gaps in public knowledge, or whether this is too optimistic a view.

We know, of course, that individuals approach everyday risks in different ways, and it is hard to describe as 'irrational' choices that are governed by direct perception and experience. It is in fact when we pause for thought that the scope for getting mixed up increases by several dimensions. The following quotation from John Evelyn on some proceedings in the Royal Society in the times of Charles II illustrates that the problem is hardly novel:

> Sir Richard Bulkeley described to us a model of a chariot he had invented, which it was not possible to overthrow in whatever uneven way it was drawn, giving us a wonderful relation of what it had performed in that kind, for ease, expedition, and safety; [however] it would not contain more than one person, was ready to take fire every ten minutes, and being placed and playing on no fewer than ten rollers, it made a most prodigious noise, almost intolerable.

He adds,

> A remedy was to be found for these inconveniences.

From a study of this passage, all the elements of a sound policy on risk regulation can be inferred, including the need for the principle of reasonable practicability and the balancing functions of a risk assessment. What can also be inferred is the habit of the human mind, in matters of risk estimation, to get, metaphorically, from the frying pan into the fire. If it is true that no activity is risk free, it must also be true that the avoidance of one species of risk inescapably involves confrontation with some other species; and this is the nature of most errors, if such there be, in people's perception. The problem, if this is it, may however be no more readily curable than most other forms of neurosis, or for that matter, schizophrenia.

It can certainly do no harm to proceed on the basis of filling gaps in knowledge, whether these are relevant or not. But it has to be borne in mind that societally regulated risk is political in character. In our document 'The Tolerability of Risk from Nuclear Power Stations', the Executive described the 'societally regulated risk', in contradiction to risks we bear ourselves or risk from natural hazards, as:

> . . . major risks whose implications are so far reaching that they have to be brought under common scrutiny even though the immediate benefits may include a better life for all. The principal characteristic of such risks is that the relevant judgements cease to be in the hands of the individuals who bear them. The risks will be shifted around – geographically, or through time, or by substitution – so that some people bear more and others less, and the benefits too may also be unevenly distributed.

Some questions, being redistributive in nature, belong to the political system; and as with other political questions we do well to recognise that there lie behind them, very often, profound arguments of a moral or ethical kind that one must assume can never be resolved. A risk of a million to one against seems very like a certainty in one of a million chances; or the disturbance of one hundred million body-cells in different individuals, depending on its nature, can be statistically predicted to produce the death of one of them; and who is to say that resistance to that piece of materiality is wrong or is irrational? Killing people is normally regarded as wrong however sneakily it is done. It depends whose rationality one prefers; no doubt a conference brooding on public acceptance of innovation, and knowing that omelettes cannot be made without breaking eggs, will be predisposed to a particular view on that.

It is perhaps worth asking to what extent different political and ethical systems react in the same way to risk – assuming of course that there is freedom to do so. It seems clear for example that in the Third World, the benefits from industrialisation strongly influence attitudes; though whether this extends to those on the receiving end of the hazards is not always so obvious.

Much nearer home, I was interested recently to hear the British Chairman of the Channel Tunnel Safety Authority point out that in the six weeks previous to his making this remark, the British and French television networks had each produced major programmes on progress with the Channel Tunnel. Neither had, as a matter of fact, concerned themselves in the least with the innovatory or technological aspects of this huge project. Indeed, each had concentrated on a single evidently very popular issue – but not the same one. The British programme had been concerned exclusively with safety. The French programme had been concerned exclusively with money. While pondering on this, I came across an interesting passage in Sir Edward Spears' book on the fall of France in 1940, about a visit prior to the catastrophe of a British parliamentary group to the rear parts of the Maginot Line. They had been discussing the obsessive concentration in their British constituencies on the blackout, to the extent that even candles could not be shown, and military convoys proceeded down roads lit by the newly invented cat's-eyes with headlights blacked to allow a pencil of light downwards. Behind the Maginot Line, however, within ten minutes' distance of the Luftwaffe, no precautions whatever were taken to darken convoy lights.

The British parliamentaries sought a view from the French military authorities and received the highly practical reply that more lives would be lost through slower movement and accidents in the darkness than through any bombs the Luftwaffe were likely to drop. Perhaps they changed their minds a few weeks afterwards. All the same, it may illustrate a quite profound difference in British and French attitudes to safety – a greater propensity on our part to refuse dramatic and visible risks, and an indifference to familiar ones? – a difference which perhaps survives today in the seemingly different attitudes of the two societies towards nuclear power?

Certainly 'the public', and its numerous sub-groups, are not a homogeneous mass with a single attitude to human life, rational or irrational. What for example is one to make of the, admittedly unusual group referred to in the following extract from *The Times* in early March 1990:

> Even hardened local residents were shocked at the weekend when they found a group of boys playing football with a severed human head. This incident added to the feeling among New Yorkers that violence is getting out of hand.

With all this in mind, let us now confront more directly the question of whether the filling of a knowledge gap is an answer to any problem of public perception of risk. In favour of this, we should recall that societal management of risk is a political problem, and contrary to all appearance, reason and knowledge play a very powerful part in the way governments come to decisions even in matters where strong prejudice abounds, and government has a very powerful influence on the way things happen; so analytic and comparative thinking on the subject of risk is not vain.

In favour of the contrary proposition, however, HSE's experience with its document on 'The Tolerability of Risks from Nuclear Power Stations', already cited (see page 17), may perhaps be quoted. This document was an attempt to communicate at the level of the intelligent sixth former what is known about nuclear risks and how these particular risks compare with others met in daily life. The word 'tolerability', which is new in the risk vocabulary, refers to the political element, and implicitly accepts that risks beyond those normally acceptable can be and are taken for societal reasons. In the words of the document, it 'refers to a willingness to live with a risk so as to secure certain benefits and in the confidence that it is being properly controlled'.

The document was written in the conviction that it is not the regulator's business to seek to persuade, nor is it the expert's business to decide for society what levels of risk are tolerable; but it is the business of both to explain the standards to which they are working, and their effects, so as to enable the public to form its view. The public can then, if it wishes, supersede the judgement of experts – operating, of necessity, through the political channel.

The document was applauded, even admired by the experts, despite its somewhat simplified presentation. The public, apart from some 2,000 good-hearted citizens, did not read it. The pressure groups regarded it as a serious threat to their own manner of reasoning, and put hired assassins on its track. Resort was had to the suggestion (put to me with incisive elegance by counsel for the Friends of the Earth at the Hinkley Point Inquiry) that we had, as it were, sneaked it past the public as a way of legitimising our (inadequate) standards. Probably the main effect of the document was to demonstrate that the regulator had, when challenged, a coherent philosophy about risk. But the effect upon filling any knowledge gap underlying the public's perception of risk could be described as remote. We may perhaps conclude that even if the filling of knowledge gaps would improve public perception, it is not an easy thing to do.

Perhaps another conclusion one might draw is that the public as a whole is not very interested in risk or even hazard, unless it impinges on them unexpectedly; and that when it does so with any force, the emotions so generated will defeat any attempt at preparation by experts. This is particularly so where a hazard touches on any of the unalterable points of reference the human race has about hazard, e.g. the wish not to die a lingering death, or that the race should not be extinguished, or children harmed. The public may well rank hazards according to their nearness or distance from such points of reference rather than according to calculation in any form; and perhaps there would be nothing irrational in that.

One might go on to speculate that, once the dog gets a bad name, as has happened to the nuclear industry, very little can redeem the situation except some very strong contrary perception of benefit – as might indeed happen through ecological or environmental threats attached to the use of alternative forms of power. This is partly for the reason that stereotypes are applied for purposes other than the ones which formed them. I was very recently for example astonished to read in a newspaper that, due to advice by HSE, half of a housing estate some distance away from a chemical plant was to be bulldozed. I asked our specialists what advice of theirs had led to this and how they had calculated the marginal risk effects with such precision as to divide the estate exactly into two; and why, if the danger was so acute, the empty houses were not retained as a blast barrier? They said that they had been asked for and given advice, but to an opposite effect. However the local authority concerned had for many years wished to demolish these houses and the risk argument though not supported by HSE had been useful in debate. The possible retention of the houses as blast barriers had indeed been considered, but the danger from rats was thought to exceed that from the chemical plant. The HSE specialists pleaded innocence also of this latter calculation, though they thought it very likely correct.

For many reasons, however, it would I think be wrong to accept the public perception of nuclear power or radiation as a paradigm for the way people generally perceive hazard and risk from newer technology. If ever a technology was formed and developed in a manner which would most excite the deepest suspicions and antagonism of the public, this would be true of nuclear. As I have suggested elsewhere, deep at the heart of the public attitude to it is the Faustian terror of people messing about with too much power. As if to demonstrate this more clearly than the laws of physics could, nuclear came to us with bombs, and with all the paraphernalia of military secrecy and tight lips; and the attitudes of those involved were directly influenced, dare one say it, by a kind of arrogance and perhaps by the feeling that they had given birth to the Queen of the Sciences. Radiation, too, is more precisely measurable and discoverable than other harms and itself carries a literary burden of death rays, mad scientists and strange adventures in space, so that suspicion, and epidemiology alike, fix instantly on it if any ground arises.

Very different, for example, is the current state of the public perception of the AIDS problem. The public fastened very quickly on to the idea that here was a most important threat to the future and, more seriously perhaps, the pleasures of the human race. This judgement may ultimately prove correct. But for the time being, the public appears to have recovered its nerve, probably due to the appearance of statistics which show that deaths within the strictly heterosexual population in the UK are so far few, and to the talk of vaccines that postpone the onset of the symptoms in those infected. This, I would suggest, is mainly due not so much to knowledge as to the strong perception of individual benefit and the desire therefore to be reassured.

But, though nuclear power may not be a paradigm, it is certainly a warning that innovators need to take to heart. They may nevertheless take encouragement by contrast from the, so far, successful introduction of some new technologies in recent years. One of these is the new science of genetic manipulation, an area where judgement can be confused by quite profound ethical questions, but where it has not – as with most other technologies – been necessary for large accidents to happen before those exploiting it consented at HSE's initiative to the imposition of strict controls.

In this, as in other areas, the attitude of the risk regulator is of great importance to the innovator. The regulator has a responsibility to the public to satisfy him or herself and them that what is being done is acceptably safe. He or she has a responsibility to the innovator to ensure that regulatory requirements are no more onerous than justified on an informed view of the hazard, with due allowance for the unknown. He or she has a duty to be known and respected by both public and innovator. HSE acknowledges as one of its final outputs, the following:

> . . . an increase in public confidence/decrease in public fear of industrial processes.

We have tried to carry out this policy in several ways; by coming forward quickly with guidance in fields such as the use of VDUs or on electronic process control which in fact offers prospects of greater safety than existing technologies but which cannot be fully exploited while many manufacturers remain afraid of the hazards, and by acting quickly and publicly, as in the case of legionnaire's disease, as soon as the symptoms of an unnecessary public scare manifest themselves. In this role, the perceived independence of the regulator of course plays a big part in fortifying public confidence.

Indeed, perhaps I have already said enough to betray my own prejudices about the public perception of risk. The approach to it has to be handled not by repairing gaps in public knowledge (though this aspect must not of course be neglected) but by creating well-justified confidence; and this includes a proper attitude to informing the public and to accepting the need for proper precaution. The main parts of the chemical industry in this country have in recent years followed a far-sighted policy in these matters, as have some others.

I am well aware that I have not so far fulfilled my promise to suggest an approach to the very difficult question of how much it is worth paying to avoid particular major hazards, though I have commented on the immense sums spent, with doubtful success in terms of public attitudes, to make nuclear power generation ultra-safe, and to reduce emissions of radioactivity to air and water to very low levels. In the nuclear area, however, the expenditure of such very large sums can be justified by reference to the potential economic cost of a very large accident. It could probably not be justified by reference to any conventional valuation of a human life unless a very large and subjective value were placed on pain, grief and suffering, and an explanation afforded why pain, grief and suffering shared by others in case of a major accident should be valued higher than that borne by individual families from any of the numerous vicissitudes of life.

When we came to the question of major accidents of lesser significance, terrible as they must be to those in any way involved, the judgement surely must be still more severe, bearing in mind the very large number of accidents that are in principle possible, and the huge prospect of vicissitude and adversity that the human race anyway encounters. At the end of the day society has to insure its own risks, and it has to spread the premiums pretty thin if those legendary parts that nothing else can reach are to be attained. It might be otherwise if the inventors had available to them the means of altogether excluding particular outcomes – such for example as two railway trains colliding; but that is surely forever impossible if speeds much higher than zero are to be practised.

That is not to say that large sums will not and should not be spent in reducing risk from major hazards; nor is it to say that work in comparing different degrees of risk and different expenditures is other than important. It is to say however that what is spent is as likely to be determined by the political forces and the need for confidence in particular projects as by calculation; and among these political forces are of course risk experts and purveyors of 'never fail' systems and zero-risk philosophies as well as politicians with constituencies to please.

A rational view, however, and here I conclude, is that the most effective expenditure of money, if the real object is the overall reduction of industrial risk, is upon reasonable precautions administered by an effective regulator, and with the prudent use of new and innovative technologies that are developed with safety in mind. Step jumps in safety are achieved not by huge uncovenanted investments, but by the systematic reduction of gross error and ineptitude, and the public knows it.

CHAPTER 3

The Role of the Media

Geoffrey Lean

Stanley Baldwin, in his impassioned attack on Beaverbrook and Rothermere at a 1931 election rally, accused them of 'aiming at power, and power without responsibility – the prerogative of the harlot through the ages'. The phrase has stuck in the public imagination ever since, and is generally assumed to have been his own, though in fact he lifted it – without attribution – from Kipling. Not all his party approved of his choice of words. The 10th Duke of Devonshire, with an earthy realism that would have appealed to today's tabloid editors, expostulated: 'Good God, that's done it. He has lost us the tarts' vote!'

The media certainly have power, though they are often shy of acknowledging it. Most people, at least initially, get most of their information on innovation and risk from the media. Furthermore the media act as two-way channels. They not only carry the information that helps to shape public attitudes, but they convey those reactions back to decision makers.

I believe that this dual process can take at least part of the credit for important action to reduce unacceptable risk. Media coverage of the hazards of lead in petrol, and the reaction of the public to them, played an important part in persuading the government first to reduce the level of the toxic metal and then to phase it out altogether. Similarly it did much to arouse public, and political, concern about the depletion of the ozone layer. For the two years after the publication of the discovery of the Antarctic ozone hole (Farman *et al.*, 1985), there was little movement in public policy in Britain; indeed, the Government – though it now chooses to obscure the fact – was holding up the process that eventually led to the Montreal Protocol, to the Vienna Convention and an international agreement to phase down the use of chlorofluorocarbons (CFCs). Policy only changed after the issue was taken up by the press in the spring of 1987.

On the other hand, the media have undoubtedly helped to promote public misconceptions. Most people appear to be worried about public exposure to radiation, and most appear to believe that the main source of it is nuclear power. Few seem to understand the overwhelming contribution from natural sources to both individual and collective doses or the dominance among man-made sources of medical exposures. Most, if asked what could best be done to minimise exposure, would advocate further controls on routine emissions from nuclear power stations, unaware that it would be far more effective to cut widespread overexposure to medical exposures and reduce radon levels in homes (Lean, 1985). There are, of course, real grounds for concern over both the cost and safety of nuclear power, but the media must take a share of the blame for the distortion of public concern over low-level radiation.

In 1987, MORI, the opinion pollsters, asked people in Britain to say who they trusted to tell the truth about nuclear power. Sixty-three respondents, interestingly, said that they trusted people living near nuclear power plants either 'a great deal' or 'a fair amount'. Television news came second with 52 per cent, followed closely by environmental groups at 50 per cent. Newspapers and magazines ranked next, but some way behind, with 33 per cent. (I would like to think that if the newspapers had been broken down into quality and popular press, the 'heavier' nationals would have scored better!) The three sources least trusted by the public in this survey were the nuclear industry (with 28 per cent), government ministers (at 25 per cent), and – last of all – senior civil servants (at 21 per cent) (MORI, 1987). In fact the media's influence is probably even more important than these results suggest, because the environmental groups overwhelmingly communicate their views through reports in newspapers and on television. The media as a whole are certainly better trusted than official sources of information, even if that trust is not particularly high. Two other surveys, carried out in the United States after the Three Mile Island accident, suggest that public trust in official sources is also low in the United States. (Goldsteen and Storr, 1982; Mitchell, 1982)

Another United States survey reflects poorly on the media. This found that 55 per cent of the respondents felt the media 'told the truth' about environmental issues, but 58 per cent perceived political bias in their coverage, and 81 per cent thought that it was selective and sensationalised in an attempt to maximise the audience (Ostman and Parker, 1986/87). (Obviously, again, I would like to think that a survey of the UK quality media would produce rather more comforting results!) There is an intriguing apparent contradiction in the figures. A substantial proportion of the 81 per cent who believed that the media were selective and sensational must nevertheless have also been among the 55 per cent who felt that the media told the truth.

This may, however, be not quite so paradoxical as it seems. The challenge for those of us who work in the media is to report honestly and accurately, and at the same time write articles that are readable and interesting enough to engage the general public. This does not mean that we have to sensationalise or be dishonest in what we choose to report. But we do have to select what we use, and write it simply. We work under very considerable constraints of time, space and newsworthiness. I will deal with the question of news values first, at some length, because this is, perhaps, the most important.

News Values

For most of the twenty years that I have specialised in the environment most newspapers have not thought the subject newsworthy. For most of the thirteen years that I have worked at *The Observer* I have been the only full-time environment specialist on a national newspaper; the present proliferation of good journalists working in the field only began in 1986 and only gathered pace in 1988. Our surveys showed consistently that there was great public interest in the subject; but it took a long time for most so-called opinion formers to pick this up.

One reason for this, I think, is that the environment was a relatively new issue; it was not around at the time when many journalists, and particularly many key executives, started in the profession and first formed their judgements as to what was 'newsworthy'. This was compounded by the fact that it had been briefly fashionable between 1972 and 1974 and then fell out of favour, partly because some of the wilder predictions, such as in the Club of Rome's Limits to Growth, were quickly discredited. So the subject was both new and passé, a fatal combination.

There are also more fundamental reasons. Environmental issues did not fall easily into the adversarial style of Britain's public affairs. They were not fought out in parliament between the Labour and Conservative parties, or on the shop floor between labour and management. Many journalists' news values were largely formed in this adversarial context, so they naturally felt that if an issue was not fought out in these forums it was neither important nor of public concern. And, it could be argued, many newspapers only became interested in the environment after it became political; certainly such events as the Prime Minister's speech to the Royal Society on 27 September 1988 and the 15 per cent vote for the Green party in the 1989 European elections had a galvanising effect on the media.

On the whole, the media are good at reporting debate and division, but poor when it comes to recording the process of reaching a consensus. They are good at reporting events, but bad at reporting processes. Thus the announcement of the 'discovery' of cold fusion created a media bonanza, whereas the long history of fusion research at Culham normally attracts very little interest. The oil spill from the Exxon Valdez was, understandably and rightly, a big media event; but there is too little coverage of the far greater amounts of oil discharged routinely each year into the oceans. This is partly because news, almost by definition, has to be new and unusual.

Similarly an air crash that kills eighty people will attract a vast amount of national media attention, but there would probably be no coverage whatsoever of eighty car crashes which each claimed a single victim. Public perception of events is very heavily weighted by their magnitude and very lightly weighted by their frequency.

How far all this is a cause, and how far an effect, of the media's news values is unclear, but it does seem that the public's threshold of boredom is low, and the media's threshold is, if anything, even lower. The media also tend to imitate each other. These two factors together partially explain why one particular type of story suddenly receives intense media attention and then dies away as abruptly as it appeared. For example, at one time attacks by Rottweilers or other dogs were being reported prominently every day. It is not that dogs all over Britain suddenly developed a fit of collective aggression; what happens is that a single particularly horrific or well-publicised case – or, even more powerfully, two or three such cases in quick succession – fuels both intense public interest and intense media concern to respond to it. Two Canadian studies chart this phenomenon showing that public concern about air pollution dropped by half as the media turned their attention elsewhere (Parlour and Schatzow, 1978; Dworkin and Pijawka, 1982).

The history of the coverage of famine in Ethiopia during the 1980s shows signs of this syndrome. The Food and Agriculture Organisation was warning about the approaching famine two years in advance; I myself had an article about it in *The Observer* in September 1983, based on their information, with specific warnings about its severity and the countries likely to be affected. Many other journalists wrote similarly over the next year or so. But the famine did not become newsworthy until the media made it so, when Michael Buerk sent his celebrated reports from Ethiopia. In this case the 'event' that triggered coverage was a media event, and once it had taken place the famine was reported exhaustively for six months. And, to return to my original point about the power of the media, governments, who had been well warned and informed of the famine but had done nothing, felt impelled to take action which prevented a far worse disaster (Independent Commission on International Humanitarian Issues, 1985). But then, after a while, the boredom threshold set in again. Serious famine is once more developing in Ethiopia, but is receiving little coverage.

Since the media are good at reporting events while poor at covering processes, they regarded environmental events as newsworthy even while the subject as a whole found little favour. Thus the Ethiopian drought was cited by Associated Press news editors as the third biggest news story of 1984, and the Bhopal disaster ranked second (Wilkins, 1987). Yet analysis of 952 print and broadcast stories on Bhopal in the two months after the accident in the US media showed that all focused on the disaster, the immediate aftermath and on clean-up measures; there was little discussion of why the plant was built, why settlements had been established so close to it, or of the economic and social forces that helped to lead to the disaster (Wilkins and Patterson,

1987). Only after the main news events subsided did issues such as these receive more attention.

Newsworthiness is also influenced by immediacy, both in timing, and in who is affected. I started my career on a small weekly paper, *The Goole Times*, and there one quickly learned that not only must Goole be affected by every story that appeared, but that its name must appear as early in the story as possible. Less happily, one study of the coverage of overseas disasters in the US media found that the death of one Western European seemed to equal that of three Eastern Europeans, nine Latin Americans, or twelve Asians, when it came to rating their importance (Adams, 1986).

Journalists have to work within the constraints of the news values of the time, but that does not mean that they should not strive to change them. Indeed I believe a reappraisal is long overdue, and may indeed be beginning to take place. The next development may well be a shift away, to some extent, from the dramatic and the immediate.

David Astor, the great former editor of *The Observer*, used to say that there are two kinds of scoop – the ordinary news scoop, and the 'scoop of interpretation' where a newspaper commands attention because its analysis is the best. *The Observer*'s great success of the 1950s and 1960s was largely based on the scoop of interpretation, and for a while that is what distinguished Sunday newspapers. This tradition was broken by another great editor, Harold Evans of *The Sunday Times*, who elevated the news scoop into an art form, established an excellent tradition of investigative reporting, and did great public service by shining a light upon dark corners the authorities hoped to keep hidden. Everyone was influenced by his success, and indeed securing a good news scoop is perhaps the most invigorating and enjoyable experience in journalism. But while I strongly believe that the investigative role of the press should continue – and indeed be strengthened, since it has grown weaker over the last decade – I am also convinced that the 'scoop of interpretation' needs to be restored to something of its former prominence.

There are some signs that this is happening. The daily *Independent* – though strangely, and disappointingly, not the Sunday – did lay more stress than its rivals on analysis, and this may have been one of the key ingredients in its success. At *The Observer* we have paid more attention to longer-term issues than many other newspapers – hence our long commitment to the environment – but we too are strengthening this aspect of our approach. In spring 1990, for example, we produced a special two-part supplement on global warming which attempted to give the reader detailed information and analysis that was not available elsewhere. We are also exploring ways of examining an issue intensively in several media at once. We have set up an environmental film-making company – called Central-Observer – with Central Television, which also has a long record of covering these issues, in association with a registered charity, the Television Trust for the Environment, who receive a third of the profits. In spring 1990 we also produced a special networked documentary on 'Children and the Environment', in association with the United Nations Children's Fund (UNICEF) and the United Nations Environment Programme to coincide with World Environment Day, when a special UN report on the subject was published (UNICEF/UNEP, 1990). There was a special feature in *The Observer*, and a specially commissioned book (Timberlake and Thomas, 1990). We hope this will be only the first of a series of such exercises.

Constraints of Production

There are other important constraints on the media besides the need to satisfy news values. Journalists usually have only hours – at best – in which to find out and digest an often complex array of facts and opinions, assess their significance, decide what weight to place on widely differing views and interpretations, and write or broadcast their material. Moreover, they have to compress the issues into a few hundred words, at the most, in simple, everyday language. This inevitably leads to simplification, snap judgements and concentration on immediate factors at the expense of others that may prove to be more important in the long term.

Furthermore, an article will go through the hands of several different people after the author has finished with it, and each is likely to make changes. The alterations will cut out superfluous words and sentences, make the material simpler where necessary, and cut it to fit its allocated space on the page or time on the air. This is necessary work, and generally improves the writing and presentation of the story, but inevitably mistakes and oversimplifications can occur. Because of time constraints, the original author of the piece will generally have no opportunity even to see it after this processing, let alone have a chance to put right any errors that have crept in. When newspapers were set in hot metal, the final cuts were made by a 'stone subeditor' (the adjective would sometimes reflect his personality as well as his position at the newspaper 'stone' where the pages were made up), who would have to read the story in reverse image, usually upside down, while deciding which lines and paragraphs to discard in order to make it fit.

Even with the best will in the world, these time and space constraints are going to produce mistakes and distortions, and I would defy anyone to guarantee otherwise in similar circumstances. And, of course, mistakes are not confined to journalists.

Once, in my extreme youth, I used to believe in the impartial, meticulously accurate scientist, motivated purely by a search for knowledge and punctilious in the reporting of his or her findings. As I got to know scientists, I realised that they were human too. I can remember one occasion early in my time at *The Observer* when we were working on an investigation with a team of leading medical specialists. For months we suffered from some of them saying how unhappy they were to be collaborating with a newspaper, even though they had approached us in the first place and we were financing their research, and there was the continual implication that we would distort their conclusions when we published them. In the end, we never printed the results – not because they were not sensational enough for us, but because they did not come up to our standards of accuracy. I still have a ten-page critique of their report, made for internal consumption. Similarly, with no expert eye and a limited number of sources, I regularly find mistakes in scientific reports and papers, even though they have been compiled by scientists of high repute, and have been refereed before publication.

Some journalists do indeed rely on poor evidence and sources, but again they are not alone. I remember being taken to task in a long letter by a scientist in a senior position in the nuclear industry who took exception to something I had written. In passing, he mentioned that there was good evidence that radiation was beneficial to humans. I thought this was worth reporting, and so, in replying, asked him to send me his sources. He sent back an airline advertisement for some radon-rich mineral baths.

Similarly some newspapers, not *The Observer*, are politically biased and facts can be distorted to suit their leanings. But is distortion to suit commercial or political interests unique to the media? Until privatisation proved otherwise, the Central Electricity Generating Board insisted that it was cheaper to generate electricity from nuclear power than from coal. Many of us will remember the antics of the relevant industries, and some scientists, over such issues as lead in petrol and asbestos exposure, not to speak of the way the National Coal Board appears to be trying to muddy the waters on global warming, after the emergence of a clear consensus among the world's leading scientists.

For myself, I have certainly made mistakes and will no doubt continue to do so, but I will not make them intentionally or maliciously. I confidently expect that my writing will annoy people, but not through distortion. Quite apart from the ethics of the matter, about which I feel strongly, this is a question of simple good sense. I intend to work in this field for the rest of my life, and I cannot afford to lose my reputation.

Part of the problem in the past has been that there were not enough good journalists whose future careers depended on earning and keeping the respect of people working in a particular field. Reporters who tackle one subject one day, another the next, do not have the same interest in maintaining this respect; television programmes which tackle a subject for several months, pronounce upon it with a great show of authority, and then never return to it, have even less. But now there is an excellent group of specialist reporters on the environment in both the press and broadcasting, who make up one of the best press corps in any field.

Besides this growing number of journalists willing to learn about the subject of the environment and take the trouble to get things right, there is an increasing number of scientists who are ready to communicate. And the government is finally getting the message, in this field, at least, that secrecy breeds fear – and is striving to make more information available. There seems to be an opportunity for a much closer relationship, between the various groups of people involved, in communicating about the environment, innovation and risk.

References

Adams W.C., 1986, 'Whose lives count? TV coverage of natural disasters', *Journal of Communication*, **36**, 105.

Dworkin, J.M. and Pijawka, K.D. 1982, 'Public concern for air quality, explaining change in Toronto', *International Journal of Environmental Studies*, **36**, 17.

Farman, J.C., Gardiner, B.G. and Sanklin, J.D., 1985, 'Large losses of total ozone in Antarctic reveal seasonal clox/nox interaction', *Nature*, **315**, 207.

Goldsteen, R. and Storr, J.K., 1982, 'The long term impact of a man-made disaster: an examination of a small town in the aftermath of the Three Mile Island accident', *Disasters*, **6**.

Independent Commission on International Humanitarian Issues (1985) *Famine, A Manmade Disaster*, Pan Books, London.

Lean, G. (ed.), 1985, *Radiation: Doses, Effects, Risks*, United Nations Environment Programme/United Nations Scientific Committee on the Effects of Atomic Radiation.

MORI, Market Opinion and Research International (1987) *Public Attitudes to Nuclear Power*, MORI, London.

Mitchell, R.C., 1982, 'Public response to a major failure of a controversial technology', in D. Sills *et al.*, *Accident at Three Mile Island: The Human Dimensions*, Westview Press, Boulder, Colorado.

Ostman, R.F. and Parker, J.L. 1986/87, 'A public's environmental information sources and evaluations of mass media', *Journal of Environmental Education*, **18**.

Parlour, J.W. and Schatzow, S., 1978, 'The mass media and public concern for environmental problems in Canada 1960/72', *International Journal of Environmental Studies*, **13**, 9.

Timberlake, L. and Thomas, L., 1990, *When the Bough Breaks. . .*, Earthscan, London.

UNICEF/UNEP, 1990, *The State of the Environment, 1990: Children and the Environment*, UNICEF/UNEP, New York and Nairobi.

Wilkins, L., 1987, *Shared Vulnerability: Media Coverage and Public Memory of the Bhopal Disaster*, Greenwood Press, Westport Connecticut.

Wilkins, L. and Patterson, P., 1987, 'Risk analysis and the construction of news', *Journal of Communication*, **37**, 52.

CHAPTER 4

Acceptance of Innovation: An Industry View of Environmental Aspects

J. C. Felton

Healthy industrial activity is the essential driving force of modern society. It provides the goods and services required by the population while the added value that it produces is the ultimate basis for all other activities – social, cultural and philanthropic. Thus it is in the interests of all that industry should flourish. That success depends on innovation, the life blood of industrial enterprise. The ability to do a little more, a little better, perhaps a little differently, a little more effectively, a little more economically, is what differentiates success from failure.

Equally, a healthy industry depends on a healthy environment, one in which industry can carry out its operations and which is acceptable to society as a whole. In this connection, an all-embracing concept is that of 'sustainable development' eloquently defined in the Brundtland Report (1987) as meeting 'the needs of the present generation without compromising the ability of future generations to meet their own needs'. Thus, industry must conduct its operations in such a way that it provides the products and services needed now, without prejudicing the production of these or other products and services in the future.

Public expectations with respect to environmental conservation are increasing. The quest for an improved quality of life requires that production of sufficient food must be coupled with an acceptable and accessible countryside; an increasing demand for consumer products must be compatible with the maintenance of clean rivers and beaches. These increasing demands, which have to be satisfied with new (and existing) products and processes, have become a major consideration in determining the public acceptance of innovation.

From the point of view of an ongoing, high-technology industry, any new product represents a significant investment and, to protect this investment, there must be sufficient assurance that the product will not only be acceptable at the time of first introduction but also for an adequate product life. Thus, environmental criteria rank high on the list of considerations within industry, just as they do in society.

This is both a challenge and an opportunity for industry, providing possibilities for innovation both by creating the need to achieve existing objectives better with new products and by creating new objectives. At the same time, industry must continually reappraise existing products and processes against current concerns and new knowledge. I will illustrate these themes mainly with reference to chemical products.

The Concerns of the Public

Environmental concerns that have received increasing attention over recent years range from global issues (the greenhouse effect; ozone layer depletion) through more regional effects (smog formation; eutrophication; ecosystem degradation) to local matters (individual site developments; dump sites; local effluents).

With regard to products, the public debate revolves around three basic areas of concern: toxicity, persistence, and less well-defined longer-term changes. The dramatic effects following major accidents keep the potential for toxicity in the public eye while the demonstration that some chemicals can persist and become widely distributed in the environment has created a concern about the occurrence of man-made chemicals independent of any demonstrated effect. This leads on to a concern for long-term effects on organisms, ecosystems and on the physical environment, areas where direct causation may be difficult to establish.

These concerns are real, and it is clearly right to minimise any adverse effect or change. However, there will always be areas of uncertainty and there can never be a total lack of risk. Thus, although any risk must be reduced as far as possible, it may well be necessary to accept some risk to achieve a particular benefit. Society as a whole has to recognise that trade-offs may be unavoidable, and to undertake this risk/benefit analysis.

The Acceptability Debate

The need for stability in use

Many applications require stable products. The dielectric fluid in an electrical transformer must stand up to the high electrostatic fields encountered for the long life expected of a transformer, forty or more years. A hydraulic fluid used in a mechanical pit prop has to withstand high pressure often at somewhat elevated temperature, in an environment where microbial contamination is possible and where fire is an ever-present risk. In both these cases, to be effective a compound must have the types of properties found in the polychlorinated biphenyl compounds (PCBs) that are the main products used in these outlets. Such properties can also give rise to adverse effects in the environment, for instance PCBs have been found to affect the reproduction of seals (Reijnders, 1986). Up to the present, reclamation and recycling facilities have not generally been in place to prevent the eventual release of these materials to the environment.

The heat-exchange gases in refrigeration systems must not only be stable for equipment lifetimes in the ten- to twenty-year range but be non-toxic and non-flammable in case of accidental release, particularly important for domestic appliances. These criteria are fulfilled by the CFCs that are extensively used for this purpose. It is now recognised, however, that these apparently benign compounds can have adverse effects on the ozone layer and contribute to potential global warming. As in the previous example, reclamation facilities have not been available although that situation is now changing.

If protection of timber from insect attack is required, again effectiveness must be ensured for periods of twenty years or more, a specification met by some chlorinated hydrocarbon insecticides. Clearly products that meet these exacting requirements are likely to persist if they are introduced into the general environment accidentally or deliberately. The significance of this persistence will depend on the quantity of the product released and the level at which any undesirable side-effects are produced.

The common feature to emerge from these examples is that persistent compounds used in high tonnages must be very carefully managed. They should not be used in outlets where their persistence is not required. If their release to the environment is integral to their use, then very careful consideration is required.

Selective toxicity

Selective toxicity is a fundamental requirement in the development of a pesticide and achievements have been impressive. Not only has selectivity been attained between mammals and target organisms to meet the need to protect man, but also between crops and weed species, and between pest and beneficial insects. These developments have utilised differences between pest and non-target species at every level to achieve effective selectivity in the field, differences in the site of toxic action, in metabolism, in spray retention, in behaviour, and in timing of the life cycle. These developments have utilised existing, and stimulated the development of new, knowledge which is now available for application in other areas.

Nevertheless, it must be recognised that, even with all the advances that have been made, total selectivity cannot be achieved. If this is expected of new innovations, then that expectation cannot be met. The reality is effective selectivity, and that can only be achieved by careful use.

Resistance in target organisms

Humans are attacked directly by disease organisms and indirectly affected by the pests and diseases of their crops, livestock and pets. These various attacks on our food and well-being are combated in various ways, designed to reduce the population of the pest. These control measures place a selection pressure on the pest species to which it responds by continually adapting to the control methods to which it is exposed, by becoming resistant. This phenomenon has been known from the early use of pesticides and more recently has also become important in the control of human disease organisms, for instance antibiotic resistance in bacteria and resistance of the malaria protozoan parasite to antimalarial drugs.

Strategies to prevent the emergence of resistance have been sought. These have proved difficult to develop. Knowledge of population dynamics and genetics has had to be integrated with agricultural practices to provide control of the pest species with minimum selection pressure to develop resistant strains. The development of the synthetic pyrethroid insecticides has been a particular stimulus. These were recognised as a particularly valuable group of products which acted selectively on pests on the one hand and not on mammals, birds, fish, and certain beneficial insects on the other. However, certain physiological resistant mechanisms originally selected for by dichlorodiphenyltrichloroethane (DDT) could also confer cross-resistance to pyrethroids. This concern led to an industry initiative, the Pyrethroid Efficacy Group (Jackson, 1989) now under the aegis of GIFAP, the international trade association for manufacturers of agricultural chemicals, which has taken a lead in developing strategies to contain this problem.

The overall conclusion to draw is that, if control over these pest and disease organisms is to be maintained, new products and procedures will be required, that is, the public needs to accept innovation. However, control practitioners for their part must earn public confidence by demonstrating that they are practising responsible management.

The Industry Response

Product stewardship

The response of industry to the uncertainties implicit in the environmental debate is to evolve methods to manage them. In the case of products an all-embracing concept is that of 'product stewardship'. This considers a product from the first development of a specification, through the research and development phases to manufacture and marketing, and beyond to the obsolescence of the product including plant shut down and decommissioning. At each stage the implications for humans and the environment are considered and it is ensured that satisfactory procedures are developed. This concept is clearly in harmony with the overall concept of sustainable development.

A key stage is the development of the original product specification, the template against which candidate products are to be assessed. This should include quantification of the properties influencing the performance of the product as well as those relating to any anticipated side-effects, thus providing a definition for an acceptable product. Existing compounds can then be assessed against this specification and new molecules designed if this is necessary. Ideally, this process requires input from other interested parties, the user of the product, the regulator and academia, thus incorporating a representative societal view.

Labelling

However expertly acceptability has been built in to a product, it may not be achieved in practice if the product is not used in the way it was intended. This is a process of education, and for the manufacturer the label is a powerful means for transferring the necessary information to their customer.

This process has been well studied by the agricultural chemical industry as in this case much information on correct application and safe handling procedures has to be presented to users world wide. Coordinated through GIFAP and in collaboration with the UN Food and Agriculture Organisation, particular attention has been paid to the pictorial representation of safe handling advice by the use of pictograms (GIFAP, 1988). It was found very difficult to establish simple pictures that would convey the correct message in different parts of the world. However, a standardised system has been evolved and is increasingly being used in practice and included in user training courses.

Labelling is included within the EC dangerous substances directive, and the discussions concerning the label 'dangerous for the environment' are of particular interest. The chief value of such a label is the influence it would have on the behaviour of users to attain maximum protection of the environment. Two schools of thought exist. One assumes that the normal behaviour of users is environmentally acceptable and that additional precautions should be necessary for only a few specially hazardous compounds – 'the label should be rare' school. By contrast, 'the label should be common' school feels that if a label 'dangerous for the environment' exists then any product not bearing it will automatically be handled without particular care.

Ideally such a conflict should be resolved by agreeing at the societal level on acceptable codes of conduct, in this case for handling products from the point of view of protecting the environment. Such codes of conduct could only become established within society if introduced as an integral part of the education process.

The Societal Dimension

However effectively industry fulfils its obligations, risk cannot be excluded altogether. While the public can rightly ask for increased reassurance there must be recognition that this has a price. Each added demand on a product is likely to make that product more difficult to identify and develop and hence more costly in terms of the use of resources. The price could be the exclusion of products that might otherwise have significant advantages and ultimately the elimination of any further advances. This is a problem for society as a whole and cannot be resolved by industry alone. In the last analysis, industry will provide what the customer requires and is prepared to pay for.

How do we reach this consensus as to what is or is not acceptable, the balance between benefit and risk? I would like to stress the need for judgement. We need to be able to make judgements, recognising that they involve balances of advantage rather than a perfect solution.

In entering the public debate, industry must first ensure that its own house is in order. They must ensure that evaluation processes are adequate and strive to make sound judgements within their own technologies. They must also represent fairly the benefits and risks of their products and processes.

Given this sound base, industry can then move into the public debate with a freedom of communication and a respect for the legitimate concerns of other parties. This should lead to an open, informed discussion.

In any debate, the narrower the viewpoint, the less fully the overall situation is appreciated. What is needed is training and familiarity in analysing the wide view so that the conflicting interests involved can be placed in their correct relationship and an optimum course chosen. This is the heavy but exciting responsibility of education in which we all have a role to play.

Acknowledgements

I thank colleagues, particularly I.J. Graham-Bryce and C. Inglesfield, for helpful discussions during the preparation of this chapter.

References

Brundtland, G.H., *et al.*, 1987, *Our common future*, World Commission on Environment and Development, Oxford

GIFAP, 1988, *Pictograms for agrochemical labels, an aid to the safe handling of pesticides*, Brussels

Jackson, G.J., 1989, 'The pyrethroid insecticides. A scientific advance for human welfare?' A symposium organised by the Pyrethroids Efficacy Group, *Pestic. Sci.*, 27(Special Issue), pp. 335–501

Reijnders, J.H., 1986, 'Reproductive failure in common seals feeding on fish from polluted coastal waters', *Nature*, 324, pp. 456–457, 418

CHAPTER 5

The Regulation of Pesticide Use: An Historical Perspective

John Sheail

Innovation is by definition about the future. Why then concern ourselves with past practices in pesticide regulation, particularly those pre-dating the Food and Environment Protection Act of 1985, which introduced a mandatory scheme to cover the use of pesticides and protect the environment? But is the statutory scheme so very different from what had gone before? It was, as one government minister stressed, an Act 'based on co-operation, persuasion, encouragement and partnership, but with scope for mandatory back-up measures'. If that was so, then a longer time perspective may still be relevant in explaining how present-day practices and perceptions came about. Through the greater availability of documentation, insights gained from the past may help to identify the issues most likely to weigh large in the minds of those whose task it is to decide the scope and availability of innovation.

For many years, chemists had sought a synthetic contact insecticide capable of being made in the laboratory, and manufactured on a large scale. The 'discovery' of DDT in the early 1940s gave a tremendous fillip to the search for a compound that could kill every insect with which it came into contact, but left unharmed the crops on which it was spread (Mellanby, 1989). In addition to the strides made in developing organochlorine and organophosphate compounds, the war years saw the development in America and Britain of synthetic plant regulators, MCPA and 2,4-D. An infrastructure was already in existence to promote these pesticides in farming and forestry (Dunlap, 1982). A new era of chemical control had begun.

The Safety of Operatives

It would be misleading to suggest that all scientists were mesmerised by what had been discovered. In a paper, published in *Atlantic Monthly*, Wigglesworth (1945) warned of how DDT might act 'like a blunderbuss discharging shot in a manner so haphazard that friend and foe alike were killed'. It was, however, the report of the official UK Committee on Health, Welfare and Safety in Non-industrial Employment (the Gower Committee) that first brought the potential side-effects of the new pesticides to public notice. Among its many recommendations, it proposed that employers should provide protective clothing, whenever poisonous sprays were being used (Secretary of State, 1949).

By mid-1950, officials in the Ministry of Agriculture were in no doubt that their Minister had to be seen to be taking some action to protect operatives. A conference of representatives of farmers, workers, manufacturers and contractors was convened, with a Deputy Secretary in the chair, 'to ventilate the question of how best to ensure the safety of the workers without depriving the agricultural industry of the advantages which the new substances could bring by the way of greater production and increased efficiency'. Three things had to be made clear. A large number of considerations must be taken into account. There was no evidence to suggest there was need for panic measures. An orderly examination of the problem was already in hand. The conference welcomed the appointment of a Working Party of leading scientists, with Professor Solly Zuckerman as its chairman (Public Record Office (PRO), MAF 132, 1). This too recommended statutory backing for those safety measures which many employers were already implementing on a voluntary basis (Working Party, 1951). The government agreed, and an Agriculture (Poisonous Substances) Bill was passed without opposition in 1952.

Residues in Food

By the time the report of the Working Party was published in 1951, there was mounting concern on two other counts, namely the possibility of people being harmed by pesticide residues in their food, and the effects of pesticides on wildlife. Zuckerman agreed that 'the consumer aspect was the more urgent and should be dealt with first'. His Working Party was reconstituted for that purpose. Whereas the study of hazards to operatives had taken six months, that of treated foods took two years. The issues were much less clear cut.

By July 1952, Zuckerman felt able to set out the main findings of the Working Party. Despite the steps already taken to minimise risks, members had become convinced that some kind of statutory registration and licensing system was required. Their draft report recommended that:

> . . . further studies should be made of the desirability of registering all firms or individuals who apply these materials by contract, or of farmers who undertake to provide a similar service to other farmers, so as to ensure that they are sufficiently aware of the proper techniques and times of application of the materials they handled.

Only a few thousand full-time contractors carried out spray-work. Their registration would not be onerous. A meeting of representatives from the relevant government departments, in May 1953, warned of how it might be much harder to justify the costs of keeping a register of the large number of farmers who sprayed their own and neighbouring holdings, particularly in fruit- and vegetable-growing areas. Whilst there could be no doubt that all users of toxic chemicals should be aware of the proper techniques and times of application, the meeting was unanimous in believing the more effective approach would be through propaganda, the manufacturers' instructions, and the advisory leaflets published by the Agricultural Departments (PRO MAF 130, 61).

Zuckerman's Working Party was even more adamant that controls should be imposed on manufacturers. No new chemicals or formulae, or existing products for which there was thought to be inadequate information, should be offered for sale until clearance was given by the Agricultural Departments, acting on the advice of an Advisory Committee.

Whilst agreeing that safeguards were needed, officials of the Ministry of Agriculture regarded statutory controls as unjustified and impracticable. The Working Party had failed to find any instances of illness resulting from eating food sprayed with weedkillers or insecticides. The area of home-grown crops sprayed with organochlorine or organophosphorus compounds (about which there was most concern) was still small, and there would continue to be no certain idea as to what chemicals, if any, had been applied to imported foodstuffs, which made up three-quarters of the national diet. Enforcement would in any case be almost impossible in view of the lack of accurate methods for determining micro-quantities of residues in plant and animal tissues. Analytical techniques were at best rudimentary.

The Home Office had found it possible to rely on a voluntary scheme of notification in respect of new poisons and, as the Horticultural Branch of the Ministry of Agriculture stressed, manufacturers of crop-protecting chemicals generally had a proper sense of responsibility. They could be 'relied upon to take precautions and issue instructions for guidance against the improper or careless use of their products'.

Zuckerman was unconvinced. Following a meeting in February 1953, an Under Secretary, J.A. Sutherland-Harris, put forward a compromise, whereby the recommendation made in the draft report for compulsory notification would be replaced by another, suggesting that:

> Departments should consider taking general enabling powers, as soon as opportunity offers, for use if experience shows that the making of statutory regulations is necessary.

The words 'as soon as opportunity offers' were important both from the political point of view, and as an indication that immediate action was not essential. At the very least the adoption of a voluntary notification scheme would allow time for the further factual studies which the Working Party had highlighted as being so obviously necessary. Members of the Working Party accepted the reworded recommendation, and, in May of that year, a meeting of representatives from the relevant government departments, with Sutherland-Harris in the chair, recommended the publication of their report (Working Party, 1953).

The Inter-departmental Advisory Committee

Representatives of the various government departments met in September 1953 to agree the outline of a Ministerial statement of policy. They emphasised how the complete safety of the public had to be the overriding consideration. If any damage was done, both the government and industry would suffer. Not only was it in the interests of both that a voluntary scheme of notification should succeed, but a much more practical approach was likely to develop if based on the free exchange of expertise and experience (PRO, MAF 130, 62).

A crucial factor was the prestige and influence of the Advisory Committee. Because few manufacturers were equipped to give advice on hazards, let alone the toxic properties of their products, officials of the Ministry of Agriculture expected the Committee, through a Secretariat provided by the Ministry, to identify what information should be sought, and to vet what was collected. The Ministry's Plant Pathology Laboratory had an obvious part to play in assessing threats to both operatives and consumers of treated foods. It already administered the Crop Products Approvals Scheme, whereby manufacturers voluntarily provided in confidence information on their various products for appraisal by the Laboratory as to their efficacy.

The main preoccupation of representatives of the Board of Trade at the inter-departmental meetings was to ensure that any notification scheme did not delay the development of new insecticides, which were of great importance for the export trade. As 'another hoop to be passed through before a new product could be marketed', the challenge would be to instil 'a feeling of confidence in consumers without hampering the industry'. It would not be easy. As officials of the Ministry of Agriculture minuted, it would be difficult for the Advisory Committee to know when, in the development of a substance, to ask for toxicity tests. On the one hand, many chemicals were examined, but few eventually marketed. On the other hand, it was necessary, because of the time taken to carry out the tests, that they should start as soon as possible. Likewise, there would be difficulties in deciding on toxicity standards, since the susceptibility of a mammal or insect might give little indication of that of other members of the same species, let alone the human being.

In the course of discussions with the Association of British Insecticide Manufacturers (ABIM), which represented 95 per cent of the companies in the horticultural field, it was emphasised that, if the information was to be provided at the earliest possible stage of product development, there had to be complete confidence that the data would not be disclosed, even inadvertently, to competitors. Representatives of ABIM urged that, whilst notification should apply to new chemicals, new formulations might only be covered where there was reason to suppose additional hazards. The proposal that the trade itself should decide what new products should be notified, and when new toxic hazards were involved, was particularly welcomed by the Board of Trade.

It was when the Ministry of Agriculture outlined the status of the Advisory Committee, namely that of an inter-departmental body that considered from the policy aspect the findings of the technical experts, that there was marked dissension. Representatives of the Ministries of Health and Food protested that they had assumed the Committee would be essentially technical, whereas the Ministry of Agriculture now appeared to want an almost entirely executive committee. This would do nothing to reassure public opinion. The decision to appoint an independent chairman had been intended to take the Committee's work as far as possible out of the orbit of Departments. In further discussions, it was agreed that, in view of the way the Committee's usual business would be based on 'mixed discussions of technical and non-technical data', there had to be a sense of partnership between the administrative and technical sides.

In February 1954, the Minister of Agriculture, Sir Thomas Dugdale, announced that the government had broadly accepted the proposals of the Working Party, and that Professor Zuckerman would become chairman of an Inter-departmental Advisory Committee 'to keep under constant review all risks that may arise from the use of toxic substances in food production and storage' (Hansard, Commons, 525, 115–6). It would be assisted by such expert sub-committees as might be necessary. At its first meeting, a month later, the Committee appointed a Technical Sub-committee, with the Director of the Plant Pathology Laboratory as chairman.

The Threat to Wildlife

The deliberations of the two Working Parties, and decision to appoint an Inter-departmental Advisory Committee, provided the vital context for considering the threat of pesticides to wildlife. In its report of 1955, the third Zuckerman Working Party found evidence of changes taking place to wildlife on an unprecedented scale. Whatever the precise role played by pesticides, there had to be an assurance that each product would do more good than harm, both in the short and longer term.

The difficulties of providing such an assurance were outlined by E. F. Edson, in a paper submitted to the Working Party on behalf of ABIM. It was difficult to imagine a problem with more facets and imponderables than that of the accidental killing of field animals by chemicals. It was not just that so few facts were available, or that advances in selective toxicity were one of the most difficult tasks humankind could set itself. So many conflicting interests were involved. Spraying contractors could quote many instances of farmers applauding the side-effects. Chemical pesticides killed considerably more vermin than game and attractive wild-bird species. Pesticides were, in any case, only one of many hazards. Not only did the weather, disease and predators take a higher toll, but a significant proportion also arose from people's own activities. As Edson pointed out, it was not unknown for spraying to be blamed for 'poor game prospects', when the real cause were poor standards of keepering (Scottish Record Office, AF 60, 89).

As expected, the Working Party recommended that membership of the Advisory Committee should be expanded to include representatives with nature conservation interests (Working Party, 1955). The consequent changes were made. The decision to include the wildlife aspect within the institutional framework of the Notification Scheme (later renamed the Pesticides Safety Precautions Scheme) had the effect of highlighting the inherent weaknesses of a self-regulatory system.

The first wildlife deaths that could conceivably be linked with the dual-purpose seed dressings to protect crops, namely the organochlorines, aldrin and dieldrin, were reported in 1956. The Advisory Committee asked its Scientific Sub-committee to appoint a Panel to investigate and make recommendations. In almost every case reported, the incidents occurred in the main planting and germinating period, and only grain-eating birds were affected. The losses, however, appeared small in relation to the large areas sown, and the Panel rejected any idea of there being a major threat to birdlife (Sheail, 1985).

Hard on the heels of every assurance that the problem was resolved or insignificant came further anecdotal evidence of exceptional mortality. It fell to the voluntary nature conservation movement, namely the British Trust for Ornithology and the Royal Society for the Protection of Birds, to provide the evidence that pesticides were having a damaging effect. Under the aegis of a Joint Committee, the studies carried out by a network of observers, supported by post-mortem analyses, succeeded in providing some quantitative data on the number and scale of incidents in 1960 and 1961 (Cramp and Conder, 1961; Cramp, Conder and Ash, 1962).

It was not only the naturalist who was appalled by what was happening. The unexplained deaths of unprecedented numbers of game birds and foxes aroused serious concern among followers of field sports. A working alliance of the utmost significance began to evolve between nature conservation and sporting interests. In April 1960, Lord Shackleton initiated the first of what were to be two major debates on the issues in the House of Lords. He spoke of how, in promoting such 'catastrophic casualties' to game and wildlife, the agricultural chemist had achieved a record that would have been the envy of the Borgias (Hansard, Lords, 229, 509–16, and 230, 807–38). Powerful reinforcement also came from an extremely unlikely source, the House of Commons' Select Committee on Estimates which, whilst scrutinising the Ministry of Agriculture, heard evidence, quite fortuitously, of how 'these exceedingly toxic chemicals' had caused the recent mortality of wildlife (Select Committee, 1961).

Rather than introduce an outright ban, an alternative proposal of the Ministry of Agriculture was eventually adopted, namely to restrict the use of aldrin, dieldrin and heptachlor seed dressings to autumn sowings where considerable damage from wheat bulb fly was to be expected. They were not to be used for grain sown in spring when natural food for wildlife was in short supply and experience indicated that the birds were much more likely to eat the seed. Drawing upon advice from the Advisory Committee, the voluntary ban was negotiated annually under the Pesticides Safety Precautions Scheme. It had been hard enough to persuade the sceptical and apathetic that the sudden losses of thousands of birds in 1960–1 had been caused by pesticides. The Nature Conservancy and nature conservation movement at large were to find it even more difficult to convince them of the more subtle, but much more serious, dangers posed by the ability of certain types of pesticide to persist in the environment and exert their effect for very long periods of time (Sheail, 1985; Moore, 1987).

Emphasis has been placed on the early part of the pesticide/wildlife story because of the way it demonstrates how a growing body of opinion was already convinced that toxic chemicals brought dangers, as well as benefits, before Rachel Carson caught the public imagination with her book, *Silent Spring*, published in Britain in February 1963. The warnings of the voluntary bodies and the Nature Conservancy had become sufficiently credible for the public to be looking, in some cases consciously, for a book which crystallised and gave expression to the doubts and misgivings which this evidence had implanted in the mind.

Conflict or Co-habitation?

Government ministers have continued to emphasise the compatability of responsible pesticide use and environmental protection. Outstanding issues have remained. One of the most serious has been the deliberate poisoning of wildlife (Cadbury, 1980). More could be done to help the user select 'environmentally-friendly' chemicals, and to know when, where and how they might be applied with least damage to the environment. Fundamental to the resolution of these and other points have been two questions, namely the degree to which voluntary codes of conduct provide sufficient safeguards and, secondly, how far the damage caused by pollutants, such as pesticides, might be regarded as acceptable.

One effect of the Food and Environment Protection Act of 1985 has been to give sharper, public focus to the difficulties of arriving at some kind of consensus as to the acceptability of pesticide-induced damage. In the case of human beings, the objective was to prevent any harm to operators and consumers but, in the case of wildlife, there was a presumption that some damage would always be incurred – it was a question rather of staying below the maximum tolerance level. Under the approvals system introduced by the Control of Pesticides Regulations of 1986 (SI Number 1510), it was possible to request further studies so as to clarify the situation and improve confidence in the ultimate decision as to whether a product might be approved for some particular use. However, as Cooke (1990) noted, it was generally agreed that 'there is no automatic formula for computing acceptability'. Requests for information could not be used as excuses for delaying difficult decisions. Fortunately, the decisions do not have to be definitive. If the monitoring of risks to vertebrate wildlife in the post-registration period reveals an unacceptable loss, changes can be made.

On the premise that it was important for public debate to be well informed and properly balanced, the Act provided for the disclosure of safety data by the manufacturer. Meanwhile, the size of the data package grew enormously. These trends, together with a burgeoning of committees as part of the widening consultative process, have placed enormous demands on both the specialist administrators and scientists involved. Whether it has made for greater 'co-habitation' remains to be seen. For some critics of pesticide use, the availability of more information may merely reinforce the view that 'the more we know about their impact, the more cause there is for concern'.

The debate, the origins of which may be discerned in the deliberations of the Zuckerman Working Parties of the 1950s, is not yet over. The issues of principle and practice, raised at the outset of the chemical revolution on Britain's farms, may be as relevant today, in determining the wise use of land and natural resources, as they were some forty years ago as Britain strove, in very different circumstances, to secure a prosperous peacetime pattern of agriculture.

Acknowledgement

Dr Arnold Cooke kindly commented on an early draft of this chapter.

References

Cadbury, C.J., 1980, *Silent death. The destruction of birds and mammals through the deliberate misuse of poisons in Britain*, Royal Society for the Protection of Birds, Sand

Carson, R., 1963, *Silent Spring*, Hamish Hamilton, London

Cooke, A.S., 1990, 'The acceptability of pesticide-induced damage to wildlife', in *Pesticide effects on terrestrial wildlife* ed. L. Somerville and C.H. Walker, Taylor and Francis, London, pp 291–306

Cramp, S. and Conder, P.J., 1961, *The deaths of birds and mammals*, Royal Society for the Protection of Birds, Sandy

Cramp, S., Conder, P.J. and Ash, J.S., 1962, *Deaths of birds and mammals from toxic chemicals*, Royal Society for the Protection of Birds, Sandy

Dunlap, T.R., 1982, *DDT: scientists, citizens and public policy*, University Press, Princeton

Mellanby, K., 1989, 'DDT in perspective', in *Progress and prospects in insect control*, BCPC Monograph, 43, pp. 3–20

Moore, N.W., 1987, *The bird of time. The science and politics of nature conservation*, University Press, Cambridge

Secretary of State for the Home Department, 1949, *Health, welfare and safety in non-industrial employment: report by a Committee of Enquiry*, Cmd 7664

Select Committee on Estimates, 1961, *The Ministry of Agriculture, Fisheries and Food. 6th Report*, Session 1960–1

Sheail, J., 1985, *Pesticides and nature conservation: the British experience, 1950–1975*, Clarendon Press, Oxford

Wigglesworth, V.B., 1945, 'DDT and the balance of nature', *Atlantic Monthly*, 176, 107–13

Working Party on Precautionary Measures against Toxic Chemicals used in Agriculture, 1951, *Toxic chemicals in agriculture*, HMSO, London

Working Party on Precautionary Measures against Toxic Chemicals used in Agriculture, 1953, *Toxic chemicals in agriculture: residues in food*, HMSO, London

Working Party on Precautionary Measures against Toxic Chemicals used in Agriculture, 1955, *Toxic chemicals in agriculture: risks to wildlife*, HMSO, London

CHAPTER 6

Pesticide Innovation and Public Acceptability: The Role of Regulation

Joyce Tait, Sylvia Brown and Susan Carr

The pace and direction of technological change is, to some extent, determined by public perception of the relative risks and benefits of new products. The trend at the moment is such that industry and regulators appear to be paying greater attention than ever before to public attitudes in deciding which new products to develop or to approve. However, during the heyday of pesticide innovation, from the late 1950s to the early 1970s, the influence of public opinion on decision making was weak. Industry's concern for the potential market for a new product (primarily determined by the attitudes of the farmers concerned) was largely undiluted by direct consideration of public attitudes. The regulatory system was primarily reactive to events, controls being introduced only after environmental or health impacts manifested themselves. From the point of view of the uptake of pesticide innovations and the resulting increases in productivity, the most important aspect of contemporary policy was the strong government support for agricultural prices throughout the developed world, which gave farmers an incentive to increase the level of pesticide inputs, secure in the knowledge that this would also increase their profits.

This chapter will explore the issues surrounding pesticide innovation from the points of view of industry, the farming community and the public. It will also discuss briefly the legacy being passed on to the emerging new biotechnology industries, as a result of past responses to these issues.

The Agrochemical Industry

One measure of the innovativeness of the agrochemical industry is the number of new chemicals being introduced to the market and, since the mid-1970s, this has been declining. It is often claimed that this problem is due to the soaring development costs of new pesticides but, in fact, both the increasing development costs and the reduction in the number of new chemicals are symptomatic of a mature industry with the following characteristics: the chemicals which are obvious or easy to discover, or to commercialise, are already on the market; satisfactory pesticides are already available for most of the common pests, so any new chemical would have to compete with existing, relatively cheap, off-patent chemicals; greater competition between companies has led to a gradual decline in the real price of pesticides; and lower profits have made it harder to fund research and development (R and D) of new chemicals.

The result of these pressures was 'rationalisation' – a spate of takeovers during the 1980s, and a reduction in the number of agrochemical companies competing in the market. There has also been a considerable amount of diversification, particularly manifested by the takeover of plant-breeding and seed distribution companies by agrochemical multinational companies.

The decline in innovativeness in the agrochemical industry was often blamed on the cost of compliance with the requirements of the regulatory systems of various countries (Green, 1976; Corbett, 1979). However, even today, this does not amount to more than about 25 per cent of the total R and D costs of a new pesticide, and any chemical with such a tight potential profit margin would be discarded for that reason alone. It is also claimed that regulatory costs have prevented the development of pesticides for small-scale, 'niche' markets, but these are not attractive markets for multinational companies, in any case, regardless of the costs of complying with regulations.

Public environmentalist pressures and the regulations to which they have given rise have, if anything, opened up opportunities for the agrochemical industries. In the absence of the bans and regulatory restrictions imposed on the organochlorine insecticides in the 1960s, it would have been very much more difficult to persuade farmers to buy the new, more expensive, more toxic and less persistent organophosphorus insecticides. The next major group of insecticides to be exploited was the synthetic pyrethroids and they were able to compete successfully with the organophosphates because of their greater safety to the operator and the environment, despite the fact that many organophosphates were by then off patent and relatively cheap.

When one considers the position of small firms attempting to set themselves up in competition with multinational companies a different picture emerges. Here, the cost of complying with regulatory systems and the delays introduced by this hurdle have completely prevented the emergence of significant numbers of small and medium-sized enterprises (SMEs) involved in the pesticide industry. Indeed, a full cost/benefit analysis may show that the competitive benefits arising to multinational companies from regulation exceed the total costs of complying with regulations.

A recent survey of various sectors of the crop-protection industry, based at the Institute of Arable Crops Research (IACR), Long Ashton, has provided some views of business strategists on regulation (Brown, 1990). When top-level managers were asked about their major concerns in strategic decision making, in aggregate, political and legal considerations generally came a poor third, after technological problems such as the development of pathogen resistance to the company's own flagship products and various other business-strategy and managerial issues. Financial and economic considerations were typically fourth and environmental or safety issues consistently attracted least attention. In individual cases, the organisational role and personal style of the manager concerned had an important influence on the detailed ordering of these issues. For example a marketing director of a large multinational company was inclined to take a more comfortable view of governments (seen as able to be influenced) than of merchants and distributors (seen as difficult to control). Among the merchants and distributors themselves, the more reputable favoured more regulation to eliminate the 'cowboys' in this increasingly competitive sector.

Also in this survey, companies with a strong R and D base tended to see competitive advantages in regulation where, for example, the withdrawal of older chemicals with undesirable properties created opportunities for their replacement by more sophisticated, and more profitable, chemicals. Large companies were also confident of their ability to make their views heard at governmental level, and also to move their operations elsewhere if necessary, in a search for a less restrictive regulatory climate.

Both anecdotal and empirical evidence therefore suggest that the directly inhibitory impact of public attitudes on pesticide innovation by multinational companies has never been more than marginal. Public attitudes have operated via the regulatory process, but their impact so far, at least on multinational companies, has been to stimulate innovation.

The Farming Community

As noted at the beginning of the chapter, the farming community, as the potential market for new pesticides, has had a very strong influence on the decision making of agrochemical companies. It is therefore relevant to ask what the farmers' attitudes to using pesticides are, and to what extent they are influenced by the public.

Many farmers do, in fact, have very mixed attitudes to pesticides. They feel that they need them in order to survive commercially, but many find them unpleasant to use and express concern about potential impacts on the environment. Often members of the farmer's immediate family are anti-pesticide, as are non-farming neighbours, while farming neighbours exert countervailing pressures to keep fields free from weeds, pests and diseases. Increasingly, the sources of crop-protection advice available to farmers are also in the business of selling pesticides.

The behavioural outcomes of these conflicting influences were investigated in a survey of brassica farmers, carried out under the voluntary system of pesticide regulation that was in operation in the UK prior to the implementation of the Food and Environment Protection Act, 1985 (Tait, 1983). It showed that:

(i) pesticide usage varied by a factor of over three for very similar levels of pest control achieved;

(ii) this variation could be correlated significantly with attitudes to factors such as effectiveness of pesticides, ease of application, the risks of crop losses, or wasting money on needless pesticide applications (and, although many farmers claimed to be concerned about wildlife damage or the health impacts of pesticides, these factors had consistently low correlations with actual pesticide usage);

(iii) recommendations by the Ministry of Agriculture, Fisheries and Food to avoid the use of DDT, dieldrin and aldrin on brassica crops, where there were less persistent alternatives, were almost entirely ignored (the use of these organochlorines was not at that time controlled by law);

(iv) even though food processors and the big supermarket chains claimed that no DDT was used on crops grown for them, the great majority of farmers supplying them were using it regularly;

(v) DDT was being used on crops exported to the Netherlands, where its use was illegal.

Among other things, this survey therefore indicated that the strongly adverse public attitudes to the use of organochlorine insecticides throughout the 1960s and '70s, backed up by the advice of the Ministry of Agriculture, Fisheries and Food (MAFF), had had little or no effect on the actual behaviour of farmers. This survey also found that farmers who had received most complaints about pesticide use in the past, from non-farming neighbours, were those who used pesticides most frequently, implying that the complaints had not caused them to change their behaviour.

Two more recent surveys of farmers and conservationists in Bedfordshire (Carr, 1988) showed a sharp divergence between the two groups in attitudes to pesticide use, farmers having a median attitude score of +11 and conservationists a score of –22 (sig. 0.001). Farmers' overall attitudes were mainly determined by factors such as 'ensuring high yields', 'increasing income', 'recovering growing costs', 'using high input systems', 'ensuring quality crops', 'keeping on top of pests, weeds and disease' and 'being restricted to carefully tested chemicals'. Conservationists, on the other hand, felt most strongly about 'leaving toxic residues', 'building up in the food chain', 'having unknown long-term effects', 'harming beneficial insects' and 'harming wildlife'. Another very striking aspect of this survey was the almost total lack of contact and communication, and hence of direct influence, between conservationists and farmers.

As part of the research based at the IACR, Long Ashton, referred to above, major farming companies and market gardeners, crop consultants and merchants whose services are sold on a 'high-tech' image were surveyed. Crop consultants described how they felt constrained to recommend chemical intervention so as 'to be seen to earn their fee'. Market gardeners, in particular, referred to external pressures for cosmetic quality, largely arising from supermarkets with monopsonistic powers, as increasing the level of pesticide use. This survey also suggested that the emerging organic farming movement may be leading to a greater polarisation of attitudes in the farming community. For example, a large, high-input, contract farming enterprise with a planned crop protection strategy was, from time to time, forced to use husbandry techniques where chemicals had failed. This was viewed as incongruent with the company image and rather undignified – they wanted better chemicals as soon as possible.

The attitudes and intentions of farmers are undoubtedly the strongest influence on the agrochemical industry and its strategic plans for innovation. Based on the survey results quoted here, there is no evidence that the anti-pesticide attitudes, often strongly expressed by some members of the public, are having any influence on the farming community and hence, indirectly, on pesticide innovation.

The Public

The 1960s saw the emergence of strongly adverse public attitudes to pesticide use when the environmentally damaging effects of the organochlorine insecticides first became apparent. The delays that took place while scientists proved beyond reasonable doubt that organochlorine insecticides were indeed the cause of population decline in a wide range of species (Sheail, 1985) were probably justified on scientific grounds. However, the apparent reluctance of government and the agrochemical industry at the time to consider the possibility of such damage contributed significantly to the lack of public trust in the chemical industry and to its current public image problems. In the Bedfordshire survey of conservationists referred to above (Carr, 1988, p 67) publicity about the harmful effects of DDT was still one of the most frequent reasons given for anti-pesticide attitudes.

Raymond Williams (quoted in the *Times Higher Education Supplement*, 22 December 89) described the characteristically English response to problems, both serious and trivial, as follows:

(i) it is not a problem;

(ii) it may well be a problem but it is being exaggerated;

(iii) it is indeed a problem but it is being badly expressed;

(iv) it is certainly a problem but it is being grossly/hysterically/obsessionally formulated;

(v) it is of course a problem but it is already well known and everything likely to solve it has already been tried;

(vi) it is a problem but it is (has become) boring.

As applied to the progress of the regulation of organochlorine insecticides in the UK, these observations are embarrassingly accurate.

Since the 1960s pesticide use has retained a fairly prominent place among the technological issues that give rise to public concern. If anything, the level of opposition has been increasing in the UK in recent years, particularly as regards residues in food and health effects on agricultural workers. It is, however, important to distinguish between two quite different types of public reaction, based respectively on the interests and the values of the protagonists.

It is usual to classify all forms of public protest related to environmental issues as a NIMBY reaction ('not in my back yard') (Gervers, 1987), the term often being used in a derogatory sense, implying that protesters have their own selfish interests at heart, rather than the wider interests of society. The other end of the spectrum of public protest, rarely specifically identified, is epitomised by the NIABY response ('not in anybody's back yard') where the protagonists believe that a particular technology or development should not be promoted under any circumstances (Tait, 1988). In the case of pesticides, a NIMBY reaction would be one where members of the public object to the use of pesticide on land adjoining their own where it might drift on to their garden, but are happy to buy blemish-free fruit and vegetables produced with the aid of pesticides next door to someone else's garden. Members of the public who adopt a NIABY stance, on the other hand, would like to see the use of pesticides diminish or disappear altogether, and they are often prepared to take actions that run counter to their own apparent interests to further their cause. The majority of the conflicts that have surfaced over the use of pesticides have been in this latter category.

It is important to understand the distinction between these two types of public response as the strategies likely to be successful in resolving conflicts based on them are very different. Conflicts of interest (NIMBY) can usually be resolved by the provision of information or compensation, or by negotiation. On the other hand, conflicts of value (NIABY) are much more difficult to resolve – information will be viewed as propaganda, negotiation as betrayal and compensation as bribery. Members of the public who approach a pesticide problem from a NIMBY standpoint are also unlikely to attempt to elevate it to a national issue, whereas those who approach it from a NIABY standpoint are likely to become organised on a national or international basis and to extend their concern to a wide range of other related issues, for example Friends of the Earth, Greenpeace, Pesticides Action Network.

Distinguishing between these two standpoints is not always straightforward. For example, the foundation of government risk-regulatory systems in scientific rationality often forces on NIABY-based groups a framework which does not allow them to express their arguments in their own terms. They may therefore adopt NIMBY or scientific/economic arguments to support their case, even though these do not reflect their true motivations, for example based on a cost/benefit analysis.

It is also a mistake to assume that these distinctions are found only in the arena of public attitudes. A classical example of a NIABY-type reaction to a pesticide problem is contained in a document from Dow Chemical Company (1980) which explains to shareholders and employees why the company chose to spend so much money defending the herbicide 2,4,5-T:

> The expedient course of action . . . would be to abandon the fight and divert these resources to the development of more profitable products which would (hopefully) be acceptable to those who oppose 2,4,5-T . . . [The] Dow strategy is based on the belief that *chemicals* [their emphasis] are the real target – the foes of 2,4,5-T are just warming up for future attacks on other pesticides.

The clues that point to this being a NIABY reaction are: (i) the fact that the company was prepared to act against its own financial interests in this matter; (ii) the extension of the conflict to other areas (all chemicals as opposed to just 2,4,5-T); and (iii) the use of emotive language such as 'foes' and 'warming up for future attacks'. Dow appears to have responded in kind to what they assumed (probably correctly) was a NIABY-based opposition, a reaction guaranteed to increase rather than defuse conflict.

In the research carried out at the IACR at Long Ashton (Brown, 1990) public opinion was taken very seriously by the members of the agrochemical industry surveyed. Business strategists in the sample expressed concern about the political implications of a 'groundswell' in public opinion. 'Getting the message across' was a commonly expressed need, and public-relations strategies included liberal incorporation of the word 'green' in publicity material, television and magazine advertising, and in children's activity stands at county agricultural shows. However, as noted above, there was little indication of such factors exerting a major direct influence on R and D decision making.

The Role of Regulation

As the evidence outlined above shows, the anti-pesticide attitudes of some sectors of the British public appear to have had little impact on the innovative decision making of multinational companies in the agrochemical industry, either directly, or through the farming community. Public attitudes are much more likely to exert an influence through the political system, resulting in more stringent regulations and the withdrawal from use of some chemicals found to have undesirable properties. However, as the evidence presented here shows, regulation has so far had a stimulating, rather than an inhibitory effect on pesticide innovation in multinational companies. Indeed, in the last years of the old voluntary Pesticides Safety Precautions Scheme (PSPS), the British Agrochemicals Association (BAA) joined the public-interest pressure groups in lobbying the government for a legally based regulatory system:

> . . . the agrochemical industry will not be satisfied that the safety objectives of the PSPS are being met until it is illegal to sell, distribute or use a pesticide in the UK without prior clearance under the PSPS. . . . It is a matter of regret that this government has chosen to exercise that power [to delay the inevitable] in respect of this urgent need for additional legislation. (BAA, 1983).

Although the costs of complying with pesticide regulations have not greatly inhibited the multinational companies, they have probably been a major factor preventing the emergence of a small-firms sector in the agrochemical industry. Since small companies would be more likely than multinationals to develop products for specialised, niche markets such as selective, and hence less environmentally damaging pesticides, regulation could therefore have inhibited innovation in such areas. It would be somewhat ironic if the end-result of the introduction of regulation designed to protect the environment has been to stifle the development of new, less environmentally harmful products.

The Food and Environment Protection Act, 1985, and its subsequent Regulations, like the previous voluntary scheme, have therefore not been particularly inhibiting of innovation by existing agrochemical companies. The main problem to have arisen so far during the implementation of the Act has been the delay of up to two years in the processing of applications for registration of new chemicals by MAFF. This is due to lack of administrative resources and the operation of a single 'queue' for both the reassessment of old products and the registration of new products. Industry representatives claim that this is delaying the entry to the market of more environmentally benign and safer products and reducing the active patent lives of new discoveries by up to two years. The opportunity cost of this delay is much more serious for a company than the cost of the various tests and paperwork needed to obtain clearance for a new pesticide. In this context it is interesting to note that BAA and the Green Alliance have recently joined forces to put pressure on MAFF to speed up the review of older products and to make public the safety information on the basis of which products are registered.

Despite such occasional coincidence of interests, the public response to pesticides is mainly of a NIABY type. This means that once one issue has been resolved (for example re-examination of the backlog of old pesticides cleared under a less-demanding regulatory regime) the attentions and energies of public-interest pressure groups will merely shift elsewhere. Currently, it appears that this NIABY legacy of pesticides, reinforced by some serious new ethical issues, is in the process of being transferred to biotechnology.

In the case of biotechnology, the concern of the agrochemical industry that adverse public attitudes will inhibit innovation appears to be more serious than it ever was for pesticides. The fear is that public pressures will act via national governments and the European Community to put in place regulations that will severely delay or restrict new developments and may inhibit some altogether. Particularly alarming to industry is the demand from some groups in the European parliament for a 'fourth regulatory hurdle'. The three regulatory hurdles that new products currently must clear are *quality*, *safety* and *efficacy*. The proposed fourth hurdle is *need* – products should be screened by government agencies for their potential social and economic impacts before being cleared for release to the market place. This suggestion would enable public-interest pressure groups that have adopted a NIABY stance on pesticides and biotechnology to express their objections in terms that are closer to their underlying motivations, rather than having to resort to scientific/economic arguments as noted above, and so it is likely to have considerable political support, at least in some countries of the EC. There is an urgent need for a better understanding of the basis of the NIABY response to particular technologies and of its transfer to new areas.

The history of pesticide regulation in the UK has been one of increasing confrontation between the industry/regulatory axis and the public. In the case of biotechnology, a serious attempt has been made to learn from the mistakes of the past. The new biotechnology industries are collaborating in the development of a proactive process of risk regulation, where risk assessment is built into the early stages of the industry itself (Tait, 1990). However, the somewhat fragile consensus underlying this process could break down if NIABY-type responses from some pressure groups cause serious delays to the commercialisation of new products, or if industry's willing collaboration in the regulatory process should appear to be motivated by considerations other than a genuine concern for the protection of the environment.

Acknowledgement

This chapter is based partly on the results of the Economic and Social Research Council/Agriculture and Food Research Council joint crop-protection research project, *Technology, Policy and Environment*.

References

BAA, 1983, *Annual Report and Handbook, 1982–3*, British Agrochemicals Association, London

Brown, S., 1990, *Technology, Policy and Environment: a Joint Crop Protection Initiative of the ESRC and AFRC*, Report to the Research Councils, Institute of Arable Crops Research, Long Ashton, Bristol

Carr, S., 1988, *Conservation on Farms: Conflicting Attitudes, Social Pressures and Behaviour*, unpublished Ph.D. thesis, Open University

Corbett, J.R., 1979, 'Technical Considerations Affecting the Discovery of New Pesticides', *Chemistry and Industry*, 22, pp. 772–82

Dow Chemical Company, 1980, *Why Defend 2,5,4-T? A Summary of Public Issues Important to the Dow Chemical Company*, 19, 16 May

Gervers, J.H., 1987, 'The NIMBY Syndrome: is it Inevitable?', *Environment*, 29(8), pp. 18–20

Green, M.B., 1976, *Pesticides, Boon or Bane?*, Paul Elek, London

Sheail, J., 1985, *Pesticides and Nature Conservation: the British Experience 1950–1975*, Oxford, Clarendon Press

Tait, J., 1983, 'Pest Control Decision Making on Brassica Crops', in ed. T.H. Coaker, *Advances in Applied Biology*, Vol. VIII, Academic Press, London, pp. 122–88.

Tait, J., 1988, 'NIMBY and NIABY: Public Perception of Biotechnology', *International Industrial Biotechnology*, 8(6), pp. 5–9.

Tait, J., 1990, *Biotechnology – Interactions between Technology, Environment and Society*, EC MONITOR/FAST Programme 1989–90, Synthesis Report No. 1, Project: Biosphere and the Economy, DG XII/H/3, March, 1990. Also published as, Tait, J., 1990, 'Environmental Risks and the Regulation of Biotechnology', in eds P. Lowe, T. Marsden and S. Whatmore, *Technological Change and the Rural Environment*, David Fulton Publishers, London, pp. 168–202.

CHAPTER 7

The Release of Genetically Modified Organisms

John E. Beringer

The rapid advances in our ability to manipulate organisms genetically that have occurred during the last twenty years have transformed humankind's ability to breed novel organisms. It is now theoretically possible to introduce genes from any species into any other species, and is practicable for many commercially-useful plants, animals and micro-organisms. Genetic manipulation, through the cloning and modification of genes, offers exciting prospects for agriculture, industry and medicine. However, before the new organisms can be exploited commercially there is a need to ensure that they will not harm the people using them or the environment.

This chapter reviews concerns raised by the release of genetically modified organisms to the environment, and discusses the regulatory procedures that are being developed in the UK to protect the environment.

Types of Release

Release can be defined simply as a situation where living organisms are not contained and cannot be recovered from the environment with absolute certainty. Such a definition depends upon an adequate definition of containment and what is meant by the term 'recovered'. For most, particularly large, organisms this appears simple because containment usually involves physical barriers to prevent the spread of the organism. However, it is almost impossible to ensure that no micro-organisms or viruses are spread by people working with them. In practice, this is not a serious problem because working conditions can be established to ensure that the numbers released are so low that there are no health or environmental problems. An example of the success of this approach is provided by public health laboratories, where the containment procedures for different pathogens are related to known risks of infection of the organisms concerned. Infection of workers under these conditions is unusual, and almost unknown in contacts or in the immediate environment of the laboratories.

Another important source of 'release' is through gene transfer in natural mating events. This is much more difficult to contain than the organisms themselves, especially for plants. For example, plants in a glasshouse are contained, but when they flower their pollen is disseminated widely unless the glasshouse has high quality filters to prevent this happening.

From the point of view of regulating releases to the environment it does not matter how they occur. Therefore, release committees have the problem, not only of assessing planned releases, but also of providing advice and guidelines to obviate problems that may occur from accidental releases.

Another problem concerning the regulation of releases is the definition of 'genetically modified'. This is, and has been, a major problem. Logic would suggest that the method for producing an organism is not relevant to the need for assessment before release; this should be solely on the nature of the organism itself. However, when genetic engineering started, a moratorium was called by some of the leading scientists until the risks had been defined adequately (Berg *et al.*, 1975). The response to the moratorium was that countries involved in genetic engineering established advisory and regulatory committees to judge proposals for genetic engineering research on the basis that the work involved gene cloning. Quite naturally, the success of this approach in containing the research and making sensible risk assessments has engendered the approach that the technology, rather than the product, is what should be regulated.

From a public perception point of view it is logical to assess organisms for release on the basis that they are genetically modified because they are 'novel', and the potential for harm from novel organisms needs to be assessed very carefully. However, there is little scientific basis for making judgements on this criterion and thus what is classified as genetically modified becomes important. Because much of the basis for regulation is the novelty of the technique, there is a major problem in deciding what is a novel technique.

There is no need to expand on the problems of definition here. What is practically important is that in the existing UK guidelines the onus is put upon people intending to release novel organisms to ask for advice if there is any doubt whether the organisms have been manipulated or not. In the absence of a completely clear definition of a genetically modified organism this allows regulation to err on the side of caution.

How Releases are Regulated

Until the spring of 1990 in the UK releases of genetically modified organisms to the environment were regulated by the Health and Safety Executive advised by the Advisory Committee on Genetic Manipulation (ACGM). The ACGM has a sub-committee (the Intentional Introduction Sub-Committee) which was responsible for developing guidelines and making assessments of release proposals. The sub-committee was responsible to ACGM which undertook the final review of the proposal. There was also a Department of the Environment committee (the Interim Advisory Committee on Introductions, IACI) which advised the Department on the ecological risks of the proposed release.

The Intentional Introduction Sub-Committee was a tripartite committee consisting of academics with ecological and genetic engineering expertise together with representatives of relevant government departments (such as the Department of the Environment, DoE, and the Ministry of Agriculture, Fisheries and Food), members of environmental bodies (such as the Nature Conservancy Council and Forestry Commission), and appointees from the Trades Union Congress and Confederation of British Industry. The Department of the Environment's committee did not have the same structure but was a committee solely comprising of experts.

The primary role of these committees was to review proposals against the general duties of the Health and Safety at Work Act 1974 and the Wildlife and Countryside Act 1981. Proposals were reviewed against the duty to undertake a risk assessment as defined in the Genetic Manipulation Regulations 1989 and judged with respect to their potential for causing significant harm to people or the environment. The Committees had a very important role in developing guidelines and mechanisms for handling proposals. The future regulatory structure will need to be compatible with EC developments in this field. There has been, and still is, close contact with developments in Brussels to ensure that unnecessary problems do not arise.

Other factors influencing regulation in the UK are the Environmental Protection Act, 1990, and the publication of a Royal Commission on Environmental Pollution report on 'The Release of Genetically Engineered Organisms to the Environment' in 1989. The Environmental Protection Act contains plans for legislation on the keeping, acquiring, importing and release of genetically modified organisms. As its title indicates, the Royal Commission Report makes a number of recommendations about how releases should be regulated. One recommendation, that has been acted upon already, was to amalgamate the Health and Safety Executive (HSE) and DoE committees. A single committee (the Advisory Committee on Releases to the Environment) served by a joint HSE and DoE secretariat has been established to advise both the HSC and HSE and the Secretary of State for the Environment.

As well as avoiding overlap the new committee will have the role of assessing releases of all types of novel organism, whether they are constructed by genetic modification, or are simply non-indigenous organisms. This is an important step towards the recognition that all types of novel organisms may have the potential to cause environmental harm.

For many types of genetically modified organisms there are existing product-based regulations, such as those for pesticides and medicines. To facilitate communication and to avoid duplication and unnecessary delays the various government departments have agreed to a single proposal form for releases. This provides one point of entry to the regulatory processes and ensures that all bodies with legal responsibilities are informed and can start to assess the proposal in their own way. Thus, for example, a biological control agent will be judged by the release committee(s) and by the MAFF Pesticide Committee at the same time without the proposer even needing to know that both Committees have a responsibility.

The Advisory Committee on Novel Foods and Processes (ACNFP), a joint MAFF/Department of Health Committee, is responsible for assessing genetically modified products for human consumption as well as non-indigenous foods being introduced into the UK.

UK Committee Procedures

These procedures reflect the working of the original Committee (IISC), and will probably be a good guide to the way that the new Committee (ACRE) will function. The basic principle for running the regulatory process is that each proposal for a release is judged on its merits; the case-by-case approach. Furthermore, because a new technology is being assessed a dialogue is established with the people making a release to ensure that there is the maximum amount of feedback to the Committee about the trials and any unexpected events during the experiments. Such a procedure works well at present because all releases to date have been of modified organisms that are being studied to learn more about the technology or the potential for such organisms to harm the environment. Presumably in the future certain categories of modified organisms will become exempt from the full rigours of the regulatory process. It is not possible at this point in time to predict when, or what types of organisms might be involved.

The present regulatory procedure establishes a requirement for applicants to submit a proposal for a release at least ninety days in advance on an approved form. This is examined by the secretariat and circulated to members of the Committee. If necessary, further information will be asked of the applicants at this stage. The proposal is circulated to Committee members who are invited to submit comments. It is also discussed at a full Committee meeting. At present, applicants are invited to the Committee meeting to be questioned about the proposal. This reduces delays and helps to establish a rapport with the Committee to aid in the communication that is needed for the Committee to learn as much as possible from each release.

The Committee discusses potential problems and usually requires clarification of a number of points. If these are relatively trivial they can be handled by post (with the Committee's approval); if not the proposal will be discussed again at another meeting after receiving further information from the applicants. At any stage the Committee can ask for advice from other experts.

Once the Committees, and the assessors from relevant government departments, are satisfied that the risks of a release are acceptable the modified proposal goes to the ACGM for final comment. Future proposals will be reviewed by the new Committee. The ACGM will continue to advise on molecular genetics and human health aspects of particular proposals. In general the laboratory-based research will have been assessed by procedures run by ACGM prior to a release proposal.

Public representation on the original Committee was restricted, as mentioned above, to nominated individuals from the Nature Conservancy Council, an Environmental Health Officer, the Forestry Commission, the TUC and CBI, and representatives of relevant government departments. The new Committee which subsumes the functions of the IISC and IACI has representation from public interest groups. Effort is presently being spent on examining ways of developing a dialogue with the public. This will be extremely important when products are available for commercial releases. If the public is not satisfied that the risks are totally acceptable they will not purchase or use the products.

Releases in the UK

These are listed in Table 7.1. All, but one, have involved experiments conducted during the period 1986–9. The exception is the modified yeast which was announced early in 1990, and has probably not yet been released. It is also an exception because it is the only one which carried some adverse publicity. From an environmental point of view this yeast, which has had its ability to utilise maltose changed from an inducible to a continuously expressed function, has no potential to become a problem. Because it is also part of a food, it was also assessed and cleared by the ACNFP. The difficulties in the handling of the public announcement of this product have shown a need for a very clear and generally acceptable system for ensuring public participation in the registration process.

There is nothing special to say about the other releases. All are experimental and most will be of value also to release committees in determining risks of future proposals. There are two anomalies in relation to practice elsewhere. The first is the inclusion of protoplast fusion in plants. This technique is not considered to be genetic engineering in the USA and thus the experiments listed can be done without regulatory approval there. The UK position to date is that it is better to cast the net too wide at this stage than to be too lax. Time will tell whether this was sensible.

The other anomaly is that of the bacterium *Rhizobium*. In this experiment, also conducted in France and Germany, a drug resistance transposon was introduced by conjugation to test for gene transfer in the environment.

Table 7.1 Releases of genetically modified organisms in the UK

Natural Environment Research Council, Oxford.
Four releases involving modified baculoviruses. These releases are part of a programme to develop safe and effective biological control agents for insect pests. They follow a sequence of experiments to develop an understanding of the effect of the manipulation on the viruses and the fate of the released viruses.
Agriculture and Food Research Council, Harpenden, Cambridge and Invergowrie.
Releases to study the effect of genetic modification and to look at gene transfer from modified potatoes and bacteria (*Rhizobium*).
Shell International Research Laboratories, Sittingbourne.
Research on the affect of an introduced gene into a potato.
ICI Seeds, Jealotts Hill.
Research, using tobacco as a model, on how a marker gene can spread via pollen.
Nickerson International Seeds, Cambridge.
The fusion of protoplasts from two oilseed rape species, to study the effect on growth and yield.
British Fermentation/Gist Brocades, Felixstowe.
Baker's yeast modified to express the enzyme maltase continuously. To be used to improve bread making. Not released at the time of writing.

The release was assessed by the IISC in the UK, despite there being no *in vitro* DNA manipulation, whereas in the other countries there was no regulatory assessment. The press in both France and Germany were very critical of the absence of regulatory approval. It is difficult to define the experiment as genetic engineering, which indicates the problem that will constantly arise in relation to the role of public opinion, the drafting of regulations and the functioning of advisory committees.

Conclusions

The genetic modification of organisms has reached a stage of development where we can expect products to appear in the marketplace within the next five years. Unusually for a new technology, regulations and assessment procedures to ensure the safety of the new products are being developed in advance of the products. This has the advantage that the technology can develop along with the regulatory framework thereby, hopefully, avoiding problems that could be foreseen and allowing the industries concerned to develop with a good understanding of the expected costs and related problems associated with the regulatory process. The price that is being paid is that of anticipating problems without a good background of use to base risk assessments upon.

Because this is a 'new' technology there is also a need to develop procedures to ensure adequate public representation in decision making. It is to be hoped that whatever framework of regulation emerges it will be one that enables adjustment to occur readily so that changes in technology and the need for public participation can be introduced when needed.

References

Berg, P., Baltimore, D., Brenner, S., Roblin, R.O., Singer, M.F., 1975, 'Asilomar conference on recombinant DNA molecules', *Science*, 188, p. 991

Royal Commission on Environmental Pollution, 1989, *The release of genetically engineered organisms to the environment*, Thirteenth Report of the Royal Commission on Environmental Pollution, HMSO, London

Watson, J.D., Tooze, J., 1987, *The DNA story*, Freeman, San Francisco

CHAPTER 8

Health Risks from Food

D. L. Georgala

The modern food and drinks industry forms one of the major sectors of manufacturing nationally. Retail sales of food and drinks in the UK have a value exceeding £70 billion per annum representing billions of different product packs and food portions per annum. This enormous output is used in many different ways.

If one looks in a shopping basket of supermarket goods, the interesting thing is how many different technologies are represented. A mixed range of products from a supermarket shelf or cabinet will be based on a number of primary preservation technologies such as canning, or freezing, or dehydration or chilling. These are the basic technologies on which major sectors of the food industry have developed worldwide.

In addition to these primary preservation technologies, a variety of other processing technologies are utilised depending on the product, including mixing, rapid extrusion, crystallisation, baking, vacuum packing, gas packing, fermentation, and pasteurising.

These technological developments have been an essential feature of the development of modern food distribution and the primary aims are:

- Overcoming seasonality of production.
- Allowing safe distribution of foods in urbanised societies.
- Allowing the development of variety and convenience in line with social change and demands.

This is only a very simple picture of the food and drinks industry and it would take some time to describe its complexity, because in fact it is a number of different interwoven industries each with its own technologies and its own market characteristics.

However, the point is that this very variety and diversity accounts for the complexity of food health issues and food risks.

Food Health Risks

The risks to health from food are very varied, and some of the more notable are listed in Table 8.1. These range from individual risks such as choking on one's food, and there are fatalities from this, through conventional food poisoning, through occasional infectious diseases (and hepatitis is just one and there are others) which are occasionally transmitted via food, and then we reach issues such as wilful poisoning of food for individual or other reasons.

Table 8.1 Some food hazards

Choking	Accidental poisoning
Residues	Viral infections
Wilful contamination	Natural toxicants
Bacterial food poisoning	Environmental contamination
Allergy and intolerance	Unbalanced diets

Many of these risks involve acute situations of poisoning or infection, but others, such as concerns about residues or nutritional imbalance may have a very long time-base and also be of a multi-factorial nature.

To put health risk from food in perspective the actual mortality from 'conventional' food poisoning in England and Wales is forty or fifty cases per annum, and the figure for 1986 is shown, together with other major causes of mortality, in Table 8.2.

Table 8.2 Deaths – England and Wales 1986

Circulatory/heart disease	278,749
'Cancers'	140,801
Motor accidents	4,835
Infectious diseases	2,470
Food poisoning	46
Other causes	154,302
Total	581,203

Source: Population Census and Survey – series DH2 no. 13, 1986

However, here one is talking about acute events. Some would argue that imbalance in our diet is a major contributing factor to much more widespread causes of illness and death and particularly in relation to coronary mortality, and also in some areas of cancer causation. These longer-term health risks from food remain controversial and subject to continuing research and raise some of the most difficult public health issues in terms of control and legislation.

Incidentally, perception of food risk is not always in line with reality. Expert views on the ranking of food risks frequently do not correspond with public perception. In the UK over the past year microbial food poisoning may have come into its right position in perception terms.

Microbial Food Poisoning

Given the wide range of health risks from food it is useful to focus in on one area of particular interest recently, i.e. the risks of microbial food-poisoning. As has been widely publicised now, recorded food-poisoning episodes have been rising strongly in the UK. The data for certain gastrointestinal infections in England and Wales are collated in Figure 8.1. These have been increasing sharply over the past decade, and are thought to be related to food sources. Note particularly the strong increase in Salmonella reports.

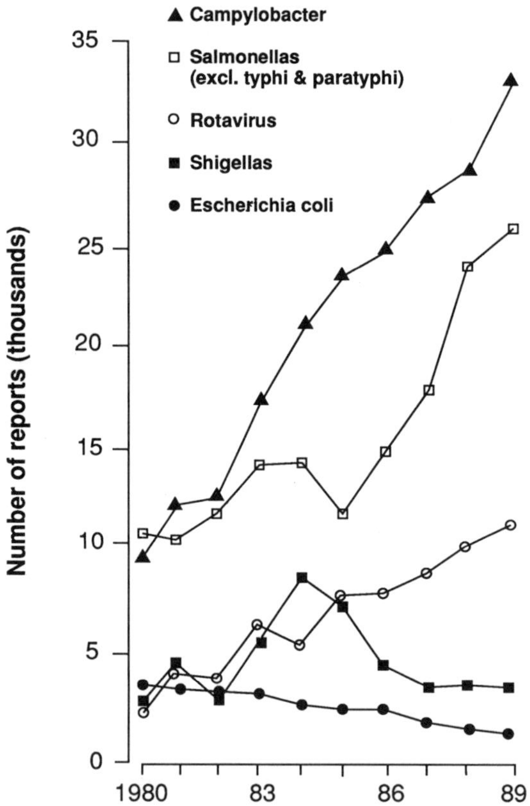

Figure 8.1 Food poisoning – England and Wales
Source: Surveillance of Gastrointestinal Infections 1980-9, PHLS Communicable Disease Surveillance Centre

These rising curves of food poisoning emphasise why there has been such strong interest over recent years in this question. The problem pre-dated the 'Famous Eggs Episode'. Within the sharp increase in Salmonella, one particular Salmonella type, *Salmonella enteritidis*, has been increasing strongly during this period. Identifying particular causes then became a major public issue, as is well known.

Incidentally, it must be emphasised that the reports mentioned above are confirmed incidents, and there are estimates that the real incidence of food poisoning is greater by a factor of ten or more per annum, i.e. even though the mortality is low a lot of people suffer every year an unpleasant illness from food.

The previous paragraphs have dealt with the more well-known food poisoning. There has also been concern about Listeria infections. The full story is not yet clear, and the role of food is not agreed by everybody, but there were 291 reports recorded in 1988.

The origins of microbial food poisoning are varied. In the first place the food poisoning pathogen has to get into the food. This can happen anywhere along the food chain, illustrated in simple form in Table 8.3. In some cases, and particularly with Salmonella, the bacteria can be established in a farm animal stock and carried through food handling and distribution into the home. In some cases farm animal infections may start from contaminated animal feeds.

Table 8.3 Food safety chain

Animal feeds
Farm animals/Fish/Plants
Factory raw materials
Food processes
Distribution and retail
Home and 'catering'

Leaving aside farm animals, contaminating bacteria can come in with vegetable or other food ingredients, from contamination from the soil or contaminated water. The very rare Botulinum poisoning can originate in this way in meat, fish and in vegetable foods. If raw materials going into food processing are sometimes contaminated or carry naturally the food-poisoning bacteria, then mistakes in processing can aggravate the problem, as touched on earlier. In addition, contamination from human or animal sources can also take place for the first time during the processing step.

Increasingly food processors are utilising HACCP systems for ensuring safe industrial processing of foods – HACCP stands for Hazard Analysis Critical Control Points. Then again distribution, retail temperature control, display time and package protection are all essential links in safety, and any breakdown here can reverse earlier good practice. Finally in the home, or in the catering establishment, even safe food can be made unsafe by cross-contamination, by poor temperature control or by incorrect or inadequate cooking.

Food Processing

Modern food processing of commercial products is a key part in the protective chain, and any breakdown in these processing steps can also aggravate or create microbial hazards. The main food-preservation systems, such as chilling, freezing, dehydration, and canning, have as a major objective the prevention of ordinary spoilage of food. But they also have a primary role in preventing the growth of food-poisoning bacteria which may occur from time to time.

Chill Preservation

The whole basis of chill storage is to reduce or stop the multiplication of micro-organisms that happen to be on food. In the first place chill storage slows down the growth of the bacteria and fungi which can spoil food by off-odours or slime formation or bad tastes. But, additionally, reducing temperature also slows or stops the growth of food-poisoning bacteria which might have happened to contaminate the food. One must always remember that the danger and severity of food poisoning increases as the actual numbers of food poisoning bacteria on a food increase. In recent times there has been growing concern about the growth of the infectious organism Listeria in chilled foods. Listeria is rare among pathogens in being able to grow at quite low temperatures, possibly below 3° C. Household refrigerators vary in their temperature profiles due to settings and way of use and, therefore, new attention is being focused on this question.

Frozen Foods

The basis of frozen storage technology is that temperatures in the region of –18 or –20° C completely inhibit the growth of micro-organisms, and also enzymic spoilage of foods. A massive industry has been built up on this technology. Provided the storage temperatures are maintained, any contaminating bacteria will be held in check. However, if a frozen food is thawed then the micro-organisms which might have been present can begin growing again. So for example, if frozen pre-cooked foods are thawed and temperature abused, and if they had originally contained food-poisoning bacteria, then the danger could be made worse because the food-poisoning bacteria could begin to grow again. Freezing does not sterilise food, it simply suspends the growth potential of food bacteria while the very low temperatures are maintained.

Dried Foods

Dehydration to produce products such as dried soup powders, dried beverages, etc. completely suppresses the growth of food-poisoning bacteria, because these bacteria need sufficient water to be able to multiply.

Once again, as with freezing, dehydration does not sterilise the product. If a dehydrated product is contaminated before it is dried, then the contaminating bacteria can well survive through dried storage, and when the product is rehydrated then, if the temperature is right, the dangerous food poisoning bacteria can multiply again and make the risk greater.

Control of water content is also important in the safe preservation of semi-preserved food such as salamis, where the product is relatively moist. However, such products usually have salt and other preservative agents which together with the reduced water content, ensure that they have extended shelf life and that food pathogens find it difficult to multiply.

Canning and Pasteurising

The canning process is over a century old. In canned foods the food product is filled into a can which is then hermetically sealed, and the whole can is given a heat treatment under pressure to ensure that the coldest part of the can receives a full heat treatment. This heat treatment is carefully calculated to ensure that it can destroy the most heat-resistant food-poisoning bacteria and particular concern is focused on the heat resistant spores of *Clostridium botulinum*. Botulism is a rare, but deadly disease, and an episode in 1988 in a mixed yogurt product will be recalled.

The heat processes given to canned foods are carefully calculated based on detailed knowledge of the death rate of Botulinum spores at different temperatures. Safety problems may arise with canned foods if there is a mistake in sterilising conditions, or if the sterilised can subsequently has a fault allowing entry of poisoning bacteria. Process control is essential and as a result such episodes are very rare.

A point really to emphasise here is that the major industries built world wide on these primary technologies have a very safe record overall. However, one can never be complacent. It must always be assumed that the raw materials used for making food products may sometimes contain food-poisoning bacteria. These bacteria may be in very low numbers and not likely to cause problems at those levels. But, any failure in the subsequent processing or in the primary preservation technologies mentioned above, can lead to growth of the food-poisoning bacteria and, therefore, to greater risk of poisoning. As mentioned earlier a key feature of food poisoning is that very low levels of contaminant in a food may occasionally cause a problem, but the risk to individuals or larger numbers of people grows strongly as the food-poisoning bacteria multiply in the food.

The basic preservation technologies which have developed over many years have played a major role in ensuring that our food is as safe as it is. One example of this is the history of the defeat of disease carried by milk and milk products as shown in Table 8.4. This dramatically illustrates the role that milk and milk products used to play in transmitting infectious

Table 8.4 Outbreaks of disease due to milk and milk products

Disease	Number of cases		
	1912–37	1938–60	1961–70
Streptococcal	5331	875	0
Diphtheria	773	37	0
Enteric fever	3229	334	0

Source: Wilson, G.S. and Miles, A. (1975) *Principles of Bacteriology, Virology and Immunity*, 6th edn, Vol. 2, p. 2684, Edward Arnold.

diseases, and the enormous benefits that have arisen from controlled heating, i.e pasteurisation, and modern dairy hygiene, which have all but eliminated the occurrence of disease transmission by milk and milk products.

Research for Improved Food Safety

Scientific and technology advances will further assist in the containment and reduction of food poisoning. In the Institute of Food Research we have three examples of research developments which will contribute.

The primary technologies for preserving foods have been mentioned. In some foods and particularly in chilled foods and some semi-preserved foods the composition of the food and its packaging also play a role in preventing food poisoning bacteria from growing. The Institute of Food Research is leading a national programme to explore in a wide range of such foods how food–poisoning bacteria grow. The aim is to understand in detail the compositional characteristics which may slow or accelerate growth of food-poisoning bacteria. These are then being put into a computer system which will enable regulatory authorities and also the food industry to check whether innovations in food can aggravate food-poisoning risks.

The risks of food poisoning begin with the contamination of food somewhere in the food chain. As mentioned earlier this can begin back on the farm or anywhere along the handling and processing of food right up to the home preparation of food. Checking for contamination, or searching for sources of contamination is made very difficult by the laborious and complicated nature of the laboratory tests for food bacteria. A number of advances have been made in recent years including the use of monoclonal antibodies to provide kits for testing for food-poisoning bacteria. The Institute has a fundamental research programme going even further, leading to the possibility of analysing and then recognising the genetic sequence specific to individual food-poisoning bacteria. Attention is being focused first on the Listeria but the principle can be applied elsewhere. The RNA sequence of Listeria is being analysed and probes are being developed which will enable these bacteria to be detected.

Used together with the polymerase chain-reaction system for amplifying genetic material in the sample, the aim is to develop principles for very sensitive tests. The importance of this cannot be over-stressed – such tests would allow detection of initial low numbers of contamination. In addition, this approach may provide a basis for a much more rapid test, e.g. within the working day, which would be a major step forward in the search and control of pathogenic contamination of the food supply.

Advanced molecular biology may open additional options for novel preservation of foods. Further research at the Institute is aimed at identifying the genes controlling natural anti-microbial substances produced by certain food bacteria. The hope is to transfer these into accepted food bacteria (such as the lactic starter bacteria), and to design the anti-microbial action so that it will specifically inhibit food-poisoning bacteria such as Listeria or Botulinum. This is a new technological approach and would raise interesting risk/benefit considerations, e.g. is a novel genetically engineered food bacterium which inhibits food-poisoning bacteria preferable to certain chemical preservatives used in foods?

In summary, our food supply, which supports large industrial populations, is basically safer than in past times. However, in recent years and due to continuing contamination in certain sectors of the food supply, coupled sometimes with lapses in process control and also due to changes in food-handling practices in catering and the home, the rise in microbial food poisoning has been worrying. With a determined effort it can be reversed. The causes are diverse, many are well known, and concerted action, enforcement, research and education will be needed.

Diet and Health

Many people suffer intolerance or food allergies. There is a growing trend towards better labelling and also databanks where dieticians can obtain reliable information about food composition, so that they can advise sufferers on how to avoid foods which cause discomfort. We must also be concerned about possible long-term health effects of natural biological molecules, particularly in some plant foods. As these are better understood, then plant breeding and also food processing can be adapted to eliminate or reduce the risk.

Here we come to the even longer-term issues of the relation between our lifetime diet from birth and its impact on our health through time. Particular attention has been focused on the role of fats in coronary disease, but there are many other factors such as fibre, minerals and vitamins. The greatest difficulty in some areas is establishing reasonable scientific agreement for action. If scientific consensus can be achieved then the health protective actions include complex policy issues, difficult education questions, and also the opportunity for new products – simple examples are non-calorific sweeteners, or non-calorific fat-like substances.

Each of these areas could require a full paper to discuss even the basic principles. So overall we have a picture of great diversity and complexity. This is also reflected in the public structures for assessment and control and policy making in relation to food safety, diet and health. There are numerous specialised committees which advise government on aspects of food safety, food additives, food microbiology, toxicology, medical aspects, etc. The fundamental law concerning food safety has now been reviewed and is embodied in the new Food Safety Act 1990.

Food Technology Innovation

Innovations in food always have to be deeply assessed for possible hazards, covering many of the risk areas mentioned earlier. The basic forms of acute risk are well known, but each episode can present unusual combinations of events which precipitate the poisoning. New food processes may reduce, or increase risks. With longer-term hazards, prediction test methods may not be reliable, and a scientific consensus may be absent. In some cases food is only one factor in complicated multi-factorial health risks, e.g. in coronary disease.

So how are innovations in food and food processing to be assessed for safety? There is no one approach. Individual defined ingredients can be tested toxicologically, and the amounts used in food and human intake can be limited by specific additive regulations.

Completely new processes or novel foods – and some biotechnology options may be such – will require complex and extensive and on-going evaluation. Toxicological, nutritional, microbiological and other evaluations may be needed, and to begin with, consumption patterns may have to be limited, and monitored, to check for safety in use.

New technological developments raise even more issues and to take one example, i.e. the likely impact of biotechnology in food processing, one can already see a number of concerns that will arise in relation to food safety assurance and also in relation to consumer perception of this new technology.

Food Biotechnology

Biotechnology by its very nature and diversity, will raise special consumer protection issues and those in food will include the following:

1 Safety of a new technology – in many cases biotechnology process

advances will be different from all previous processes. The safety assessment of new processes will have to be done for each new biotechnology process, and there may be relatively little previous data to build on in terms of safety evaluation.

2 Testing of new additives – the safety testing of new food additives is complex and varies depending on the nature of the ingredient. New biological ingredients from biotechnology may appear inherently safe, but because of lack of information doubts may be raised about their overall safety, and the proving of absolute safety may be extremely difficult.

3 Ethics of biotechnology – there are differing views on the ethics of genetically altered living systems, and this may influence consumer perceptions. In addition, national and international debate about the safety of release of genetically altered living systems may well slow the application of biotechnology in foods, including the use of altered food micro-organisms.

4 Biotechnology diversity – there is no one single biotechnology, and the approach to regulation of various different biotechnologies will prove difficult for regulatory bodies. For example, the conceptual and enforcement approach to the control of novel enzyme bioprocessing may be quite different from that required for the assessment and control of genetically engineered food micro-organisms. The conceptual base for covering the range of possible biotechnology processes is still being developed.

5 Designation and labelling – due to many different ways in which biotechnology products may be brought forward in the marketplace, there could be special labelling issues of biotechnology food products, e.g. where ingredients are made by novel technology such as plant tissue culture of the same species, or where a food ingredient is made in another species by genetic insertion of the required capability in a host, such as a fermentation yeast.

Perceptions by consumers will also depend on the information they have available about risks and possible benefits from this new technology and everything possible should be done to raise the quality of discussion and debate on this question.

Conclusion

In conclusion it is necessary again to emphasise the enormous diversity of food varieties and underlying technologies, the variety of uses and opportunities for contamination and risks to arise. Risk assessment in foods is never on a stable basis as conditions of supply and usage are changing constantly. Although the acute risks are of low frequency, longer-term concerns about links between diet and health have substantial national and international policy implications.

In the longer term new technologies and science advances will help in raising the safety status of foods, e.g. the improved rapid testing systems for contaminants, while on the other hand new technologies for processing may raise consumer concerns and also add further complications to the regulation of food safety assurance questions.

CHAPTER 9

Public Acceptance of Innovation

Derek C. Burke

This chapter, and the book as a whole, is about innovation, and innovation means change. Obviously we live in a changing world, and one thing of which we can be certain is that the future will not be like the past; as Charles Handy has said 'the status quo is not an option'. This is certainly true of the universities at this moment – as we face a new funding system from the University Funding Council, as we look at the changing demography of those who will be coming into higher education, and as we face new needs from society and from our region. It's not surprising therefore that a good part of my job is managing change within this institution, explaining why change is needed, persuading people to accept it, and solving the hundreds of problems that come with change of any sort.

Technology is one of the drivers for such change, and the sort of work that I have spent most of my life doing, working in the area between biochemistry and biology, has been one that has seen huge technological changes in my lifetime. I went to work at Yale in the autumn of 1953 after completing a Ph.D. in this country, and there on the boat was a young man who rather modestly explained that they had published a note in *Nature* that summer, because he said they had been doing some interesting model building. That paper ended up with a sentence to the effect that the authors had noticed that their model for the structure of DNA had implications for its biological function. The person was Jim Watson and the structure they proposed was the now-famous double helix. That was only 1953 and since that time we have seen the development of the whole of molecular biology, and more recently – since the mid- to late-1970s – the growth of biotechnology. Gene cloning, DNA sequencing, the expression of foreign proteins in a wide variety of cells, reverse genetics, protein engineering are all now part of the wide range of techniques we call biotechnology. Biotechnology has not only illuminated our understanding of how the world works, but it has also spawned a new industry.

As a consequence we, that is the many scientists who have grown up during this period, have become used to change, indeed have promoted change because change brought novelty and novelty brought excitement.

Molecular biologists are not modest people and though that's not always an attractive trait – and it can make them quite difficult to manage – their arrogance comes of a confidence that the problems they can identify are solvable and solvable rapidly. You may remember that the Athenians are described in the New Testament as people who 'spent their time doing nothing but talking about and listening to the latest ideas' (Acts 17, v. 21). This from the context is clearly a criticism, but I thought, as I grew up, what an attractive way to live it was! Those of you who have worked in laboratories know the excitement of finding a special paper in a new journal, rushing in to tell your colleagues and then thinking of the next experiment to do. It's fun, it's stimulating and it's been extremely effective.

However this is not to suggest that we should change everything, nor that resistance to change is inappropriate. People are resistant to changes in their lifestyle; we know from our experience here in Norfolk how slow people are to accept the idea of change whether it be building a new university residence or a new road. They don't want their world changed. More importantly, there are many things that we want to hold on to from the past; our cultural heritage, our values, our religious faith are all part of an inheritance which we value, and one of the roles of a university is to preserve those values for future generations.

There are therefore forces working in different directions, forces for and against change, and upon reflection it seemed to me that I could distinguish at least three strands within this tension. The first comes from science itself and is our growing understanding of the world we live in as a closed system; changes we bring about in one part of its complex interacting ecosystem inevitably have effects elsewhere. Jim Lovelock's Gaia hypothesis may or may not be correct but it has helped us to think in a new way about the planet on which we live, and we've all, public and scientists alike, become more sensitive to these issues. This is a fairly recent development; I was in my middle twenties when Rachel Carson's book *Silent Spring*, already mentioned (Chapter 5), appeared, perhaps among the first to draw attention to the changes that were being brought about. So we've changed our attitude to the way the world works, and that's certainly one reason to reflect upon, and sometimes to resist, proposed changes.

There is a second set of issues centring round the area of risk and value. These terms are used differently by scientists and those outside science: risk for somebody involved in genetic engineering has rather a precise meaning and, by a series of techniques developed over the last few years, can be expressed as a number which is the risk of some series of events leading to an undesired endpoint taking place. However risk means very different things to people outside science, and almost by definition it is imprecise, threatening and something to be avoided; certainly something not quantified. Value too means different things; value for a working scientist often means novelty, discovering something new and exciting; value for a company means profit or if it's a starter company just plain survival; value for the customer means something else again, maybe convenience, choice or price, increasingly now safety. This is what he or she will look for; perhaps what we should be concerned about in genetic manipulation is thinking how we can deliver the sort of value that the customer wants through our science.

Thirdly there is a more deep-seated resistance to technology which springs from what I regard as a pantheistic view of the world. Let me quote you something from a document sent to me by the Farm and Food Society:

> I think an understanding of nature lies beyond
> the reach of human intelligence.
>
> Masanobu Fukuaka

Foreword

> To the biologist it is perhaps clearer than to anyone else that all life is one. The study of living things blurs the outlines between different life forms, and the very fact that we can now split and interchange genes suggests that the same force pulses through all living matter. Not only in a physical sense is there a relationship, but Teilhard de Chardin maintained that 'psychic activity, which manifests itself fully in human thought, must also prolong itself backwards to the point where the roots of matter disappear from view'. An arresting concept, especially when contemplating the possibilities opened up by the power which now appears to be in our hands. We are on the threshold of the immense drama of man attempting to take over from nature the essence of life and transmuting it to meet his own convenience. It remains to be seen how successful this audacious experiment proves. Like so many discoveries, gene manipulation can so easily backfire, and has indeed already done so.

This is not a view that I agree with; I place myself firmly in the Judaeo-Christian tradition with an immanent and transcendent God. However this view of the relationship of the world to the divine that I have quoted treats science as hostile; an intruder best resisted, certainly not something that can help.

So we, who are responsible for change and wish to see the results of our new technologies brought into public use for the good of society, face substantial resistance and I don't think that we have yet got the systems in place which will meet these concerns and also enable us to deliver a product of value to the public.

Let me speak briefly about a couple of case studies. The first is food irradiation. I am not an expert but, reading the evidence and excluding the situation in which food irradiation is postulated as being used to cover up damage due to food spoilage – not a situation I think is desirable or likely, I see food irradiation as a helpful addition to the range of techniques that we use to deliver safe food to the general public. However it is not seen by the public like that and that largely because a vigorous campaign has been conducted against food irradiation. It is not helped by the use of the word irradiation, with its image of nuclear reactors and nuclear damage, but the argument is frankly not rational. Let me quote from a review of a book called *Biology of Food Irradiation* by David R. Murray in *Nature* (Diehl, 1990):

> The section on irradiation of spices should be of particular interest, for example, because in countries where radiation processing is practised on an industrial scale, treatment of spices to improve their hygienic quality constitutes the most important use of this technology. The reader of this book learns nothing of this. Of the 30 or so published research papers on spice irradiation, Murray mentions only two. He claims that irradiated spices lose flavour and that their treatment at the maximum allowed dose fails to reduce the bacterial contamination to a level acceptable to the meat-processing industry. A reference follows, but I found nothing in that source to support the claim.

This is not the language of reasoned argument; this is a crusade. What went wrong, why are people reacting like this? Did we just fail to explain it clearly enough in scientific language? I don't think so; there is something much more deep-seated behind this resistance which I only partially understand. Certainly the image of the scientist has become tarnished; the man in the white coat has become the enemy, the expert committee is no longer trusted. This troubles me deeply. We must try and tease out what the problems are and how we can explain our position, not that we shall win over the crusaders, but we must explain our position to the reasonable person in the general public who is so far listening more to the crusaders than to ourselves.

A second example is the case of the modified yeast which has recently been approved for use (see also Chapter 7). The production of this material was described in an article in *Nature* (Aldhous, 1990):

> No genetic material has been added to the yeast from another species. The maltose permease and maltase genes from the yeast were combined with new promoters from another strain of the same species, *Saccharomyces cerevisiae*. The genes and promoters were then spliced back into the yeast genome together with short pieces of 'synthetic' DNA. This DNA contains 'stop' codons, to minimise the chance of any hybrid proteins being produced. The end result is a yeast that should take up and digest maltose more efficiently and release carbon dioxide, which makes bread rise, more quickly.

And it was approved after careful consideration as further described in the *Nature* article:

> In its evaluation, the Advisory Committee on Novel Foods and Processes (ACNFP) says it decided that the yeast was safe mainly because the genetic manipulation was within a single species. For the same reason, and following the advice of the government's Food Advisory Committee, bread made using the yeast will not have to carry a label indicating genetic manipulation.

The Ministry put out a brief press release but the reactions of the press were quite hostile. *The Guardian* ran a piece on 3 March 1990, headed, 'Taking a rise out of mutant bread' which was not sympathetic. *The Independent* ('Man-made yeast raises temperatures', 17 March 1990) was again critical of the process. It strikes me as significant that Tom Wilkie in *The Independent* used the word 'man-made' in what I regard as a pejorative sense: man-made as opposed to natural, man-made being bad and natural being good? The actual changes that were made in order to produce this mutant yeast were minor and controlled; minor in that they might easily have happened by recombination between two strains of yeasts quite naturally in the wild, controlled in that the exact nature of the change was known and was more limited and better understood than would have been a natural mutation. But because it was done by people in white coats it's treated with hostility.

So here we have a situation in which a minimal alteration to a very safe organism is treated with hostility. That's very worrying and tells us that we don't have the proper connections made between the approval processes and the general public yet. The expert committees are doing their jobs well, and we may learn how to understand better, at an expert level, what the public means by risk and value, and take that into our consideration – as we are beginning to take into consideration other ethical isses that arise on these expert committees. But somehow or other we are simply not meeting the concerns of the people in the street and the supermarket; we are not seen to be consulting, but rather to be arrogant and overbearing. We've got to make some changes before we lose a technology, for I'm certainly not prepared to spend the rest of my working life approving new technology which the public refuses to accept.

References

Aldhous, P., 1990, *Nature*, 344, p186
Diehl, J.F., 1990, *Nature*, 344, p207

CHAPTER 10

Radiation Protection Standards: How are They Set and are They Adequate?

Roger H. Clarke

Within twelve months of the discovery of X-rays in 1895, it was reported that large doses of radiation were harmful to living tissues. The first radiation protection standards were set to avoid the early effects of acute irradiation. By the 1950s, evidence was mounting for late effects – mainly a small excess of cancers – in irradiated populations. In the late 1980s, sufficient human epidemiological data had been accumulated to allow assessment of carcinogenic radiation risks following the delivery of moderately high doses. However, workers and the public are exposed to lower doses and dose-rates than the groups from whom data are available so that risks have to be estimated for protection purposes. Even having decided on a risk per unit dose, to set dose limits, an acceptable or tolerable maximum risk must be established for both workers and the public.

This chapter presents the dose/risk relationships and discusses the issues involved in setting dose limits on risk grounds. The applicability of the limits so developed can be judged by reference to levels of radiation exposure of both workers and the public from man-made and natural sources of radiation. Conclusions are drawn about the present protection standards and areas for further research identified.

The risk estimates for radiation exposure have increased in recent years – at least for populations exposed to high doses, but the International Commission on Radiological Protection (ICRP) has yet to finalise its recommendations on how protection philosophy and practice should be amended. The result is that in various countries some interim actions are being taken, while awaiting those recommendations. This paper compares the position taken within the UK with the draft proposals just released by ICRP for consultation and outlines the reasoning behind both. The resulting dose limits are viewed in terms of their acceptability in absolute risk terms and in comparison with the variation in exposure from natural background radiation in the UK.

First a brief review of the history of the philosophy of radiation protection. The early recommendations were concerned with prevention of non-stochastic effects. These were initially defined qualitatively but when a system of dosimetry became available, quantitative dose limits could be defined. In 1934 recommendations were made implying the concept of a safe threshold – 'under satisfactory working conditions a person in normal health can tolerate exposure to X-rays to an extent of about 0.2 roentgens per day' (*British Journal of Radiology*, 1934) – about ten times the present dose limit. The tolerance idea continued and in 1951 ICRP stated 'the figure of 1 r per week seems very close to the probable threshold for adverse effects' which led to a proposed 0.3 r per week for radiation such as γ– or X-rays, with a low rate of energy transfer from the radiation to living tissue (*British Journal of Radiology*, 1951). In considering neutrons and alpha-particles, it was stated that 'anaemia and bone damage appear to have a threshold at 1 Ci Ra-226'. By 1954 the threshold was rejected and 'maximum permissible doses were such as to involve a risk which is small compared with other hazards in life' and 'since no radiation level higher than natural background can be regarded as completely "safe", the problem is to choose a practical level that, in the light of present knowledge, involves a negligible risk' (ICRP, 1955). The change was brought about by the emerging epidemiological evidence of excess malignancies amongst American radiologists and the first indication of excess leukaemias in the survivors of the atomic bombings at Hiroshima and Nagasaki.

The problem since then has become one of limiting the probability of harm and much of what has subsequently developed has been related to the estimation of that probability and the decision as to what level of implied risk is acceptable. The problem arises in both working and public environments and both issues will be dealt with in turn.

Changes in Risk Estimates

It has been becoming evident that the estimates of risk from exposure to ionising radiation are increasing. In 1987, ICRP felt the need to make a statement about risk figures following its Como meeting (ICPR, 1987), and within the UK the National Radiological Protection Board (NRPB) gave advice to government departments and regulatory agencies to consider the implications of the change in risk estimates for dose limits (NRPB, 1987). The major addition to present knowledge is the publication of the report by the 1988 United Nations Scientific Committee on the Effects of Atomic Radiation (UNSCEAR) which contains an assessment of human carcinogenic risks following irradiation (UNSCEAR, 1988).

There are several population groups who were exposed in the past to high doses of radiation and from which estimates of risk can be made, including patients treated for ankylosing spondylitis and other medical conditions, or those exposed at work. Nonetheless, the single most important source of information is from the survivors of the Hiroshima and Nagasaki atomic bombs.

Risk estimates depend largely on the latest Japanese data which have been updated for two main reasons. First, the dosimetry has been revised to take account of the high humidity in the air over the cities which substantially reduced the neutron dose, compared with the earlier values based on measurements in the dry air of the Nevada Desert. Improved estimates have also been made of weapon yield, and of tissue and organ doses, allowing for the shielding provided by buildings. In fact, the overall changes in dosimetry change the risk estimates by only 5 per cent for all solid tumours and by 50 per cent for leukaemia.

The second change is that the number of excess cancers in the population has increased since UNSCEAR last reported in 1977 due to the longer time for epidemiological follow-up to 1985. Lifetime cancer experience is not yet available for any of the large epidemiological studies. Therefore, to project the overall cancer risk for an exposed population, it is necessary to use mathematical models that extrapolate, over time, data based only on a limited period of the lives of the individuals. The two projection models that have been used in the past are:

(a) the additive model which postulates that the annual risk of cancer arises after a period of latency and then remains constant over time;

(b) the multiplicative model in which the time distribution of the excess risk follows the same pattern as the time distribution of natural cancers, ie, the excess (after latency) is given by a constant multiplying factor applied to the age-dependent incidence of natural cancers in the population.

The data now available provide a deeper insight into the applicability of the two models and UNSCEAR states in its 1988 report that recent findings in Japan suggest the multiplicative risk projection model is the more likely, at least for some of the most common cancer types (UNSCEAR, 1988). NRPB and ICRP have adopted the multiplicative model for estimating lifetime risks of most solid cancers.

The use of the multiplicative (relative) risk model implies that for the majority of solid cancers there is an increasing risk with time after exposure, following the increase in natural incidence with age. This is shown in Figure 10.1 in comparison with results for an absolute risk model. There are indications, at least in some groups exposed to radiation, that the excess risk of cancer starts to decline many years after exposure. This has been well documented for leukaemia, but has also been observed at long times after exposure in the case of solid cancers for the ankylosing spondylitis patients, the uranium miners and possibly the children irradiated for enlarged thymus. These results suggest that in the Japanese survivors, where there has been no reduction in relative risk so far, the excess risk may ultimately decrease with time. The multiplicative projection model applied over a lifetime could result in an overestimate of the cancer risks. However

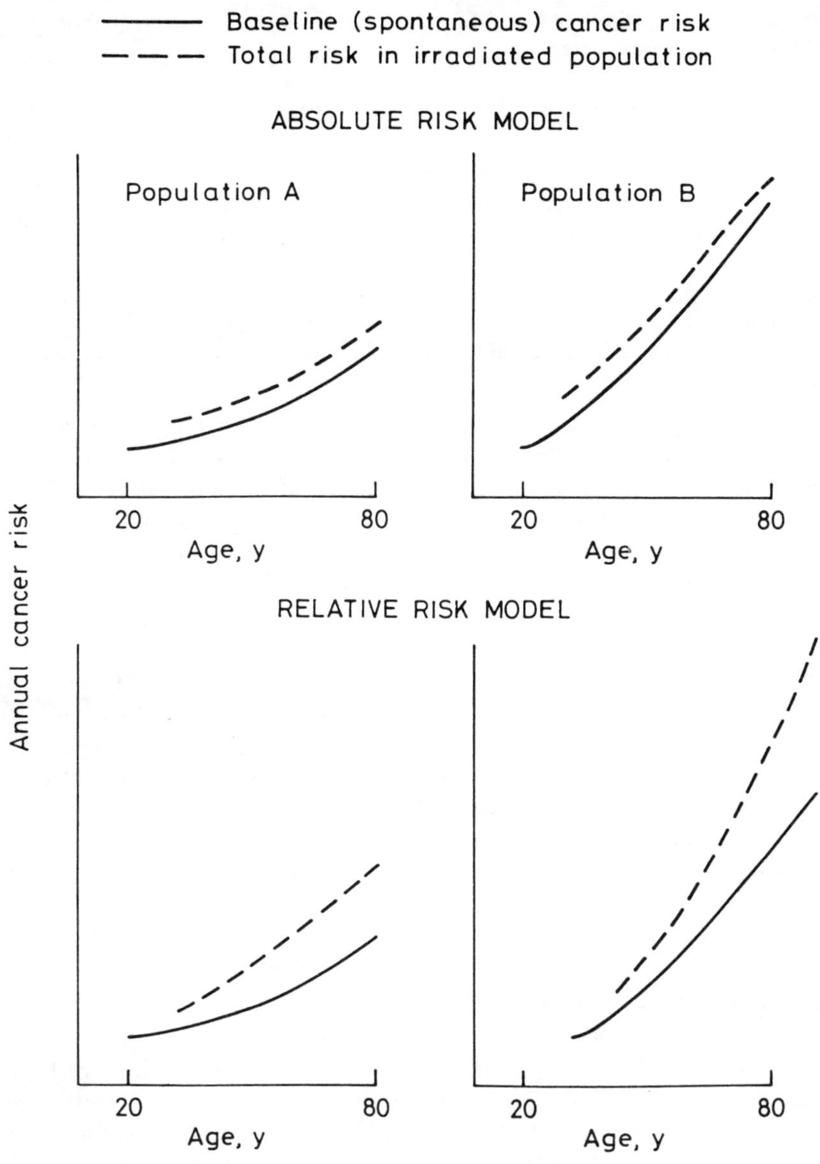

Figure 10.1 Radiation-induced cancer risk under absolute and relative risk models for two populations with different spontaneous cancer incidences

Source: National Radiological Protection Board, Report R226

populations other than the Japanese survivors, such as the tinea capitis patients and the North American breast cancer studies, also demonstrate constant relative risk.

Table 10.1 Projected lifetime risks of fatal cancer following whole-body exposure to high doses at high dose-rate
(10^{-2} Gy^{-1})

	Population	
	Working (18–64)	All ages
UNSCEAR 1977	2.5	2.5
UNSCEAR 1988	7	11
ICRP 1990	8	10

For a population of all ages, the 1988 UNSCEAR report derives a fatal cancer risk following whole body exposure at high dose and high dose-rate estimated, using the multiplicative model, at 11 per cent Gy^{-1}. For a population of working age the risks are 7 per cent Gy^{-1}. This is compared in Table 10.1 with the Committee's 1977 assessment of 2.5 per cent Gy^{-1} using the additive model. The excess deaths that have actually occurred in the Hiroshima and Nagasaki Survivors to the end of 1985 give a risk of about 4 per cent Gy^{-1} – 340 extra deaths amongst 76,000 people for whom individual dose histories have been calculated to give 9,000 man Sv.

The 1990 proposals of ICRP, based on an analysis of the Japanese data and transferring to Western population base line cancer rates, give a fatal cancer risk factor of 10 per cent Sv^{-1} (Sv = Sieverts) for a general population and 8 per cent Sv^{-1} for a workforce for high doses of low LET radiation.

Extrapolation to Low Doses and Low Dose Rates

In 1977, UNSCEAR pointed to uncertainties in two directions: the value derived from high doses was an underestimate because no projection had been made into the future, but it was also an overestimate in the sense that the risk per unit dose at low doses and low dose-rates was believed to be lower than the estimates at high doses and high dose-rates. Extrapolation into the future is still uncertain because about two-thirds of the Japanese survivors are still alive and the additional cancer risk has still to be expressed. The exception is the risk of leukaemia which appears more certain as nearly all the excess now seems to have occurred. The problem of low doses and low dose-rates remains.

Table 10.2 Summary of dose rate effectiveness factors

	DREF
UNSCEAR 1977	2
UNSCEAR 1986	up to 5
UNSCEAR 1988	2–10
NRPB 1989	3
ICRP 1990	2

The assumption normally made for radiation protection purposes is that the risk of radiation-induced cancer is proportional to the dose, without threshold. The latest UNSCEAR report does show that excess cancers are now seen, for the first time statistically significant at the 95 per cent level, in the Japanese at doses down to 0.2 Gy. As longer times for follow-up are used, excesses appear to be seen at lower doses. It is difficult to derive a dose/response relationship from the human data, but leukaemia appears to follow a linear quadratic form ($E=aD+bD^2$) in the Japanese. Solid tumours, while more linear, can also fit a linear quadratic.

There is, however, information from animal studies which indicated that the induction of radiogenic cancer by X– or γ–radiation depends on the dose-rate, and a dose-rate effectiveness factor (DREF) has been used by both UNSCEAR and ICRP to estimate the risk at low doses and dose-rates. At low dose-rates the numbers of cancers induced per Gy are lower by a factor of between two and ten than they are per Gy at high dose-rates. In 1977 UNSCEAR used a DREF of two to obtain the 1.25 per cent Sv^{-1} in ICRP Publication 26. Dose-response data for cancer induction have most recently been reviewed in detail by UNSCEAR (1986). It was concluded that for many cancers the assumption of a linear response when extrapolating from information at high doses and dose-rates could over-estimate risks at low doses and low dose-rates by a factor of up to five. In the 1988 UNSCEAR report a DREF of between two and ten is quoted, but they make no recommendations.

The majority of the available *animal* data indicate a DREF of between two and four for the induction of cancer at low dose-rates compared with that calculated at high dose-rates. A review of the available data by Stather *et al.* (1988) has concluded that a DREF of three can be substantiated for all cancers including breast cancer for which a DREF of two is judged to be more appropriate. The current proposals of ICRP are for a DREF of 2, based firstly on the Japanese leukaemia data and secondly on the theoretical calculation that if the a and b terms of the linear quadratic are equally important at a dose of about 1 Gy, then at doses up to about 0.1 Gy, the effect is half that at 1 Gy. A summary of DREFs is given in Table 10.2.

Table 10.3 Risk factors for protection purposes (10^{-2} Sv^{-1})

	Total population		Working population	
	NRPB	ICRP	NRPB	ICRP
Fatal cancers	4.5	5	3.4	4
Hereditary defects	0.8	1	0.8	0.6
Non-fatal cancers	0.4*	1.5	0.3*	1.2
Sum	5.7	7.5	4.5	6†

*Non-fatal cancers are weighted as being 5 times less severe than fatal cancers
†Rounded

Risk Factors for Radiological Protection

The risks of radiation-induced cancer have been estimated by NRPB in a UK population of all ages and both sexes. The full description of the derivation of the risks is given by Stather *et al.* (1988). The total lifetime fatal cancer risk in the UK population following whole body irradiation at low dose-rates is 4.5 per cent Sv^{-1}, over three times the ICRP figure from 1977. The fatal cancer risk estimated for a UK working population (ages 18–64 years, both sexes) exposed to whole body radiation at low dose-rates is 3.4 per cent Sv^{-1}, nearly three times the 1977 ICRP figure. The new 1990 ICRP figures are 5 and 4 per cent Sv^{-1} respectively for a population of all ages and for workers.

For an overall assessment of the detriment from irradiation of a population, information is required on cancer incidence as well as fatalities and on the hereditary defects induced in future generations.

In order to set an effective risk factor, NRPB added an allowance of 0.8×10^{-2} Sv^{-1} for serious hereditary defects in all future generations, rather than just the defects in the first two generations included by ICRP in 1977. The NRPB estimate of total risk for non-fatal cancers assumes total cancers arise at a rate of about 1.5 times the fatality rates, i.e. one-third of cancers are 'curable', but non-fatal cancers were weighted by 0.2 to give them only a fraction of the importance of a fatality. Summing these health effects gives an equivalent risk rate shown in Table 10.3 of 4.5 per cent Sv^{-1} for workers and 5.7 per cent Sv^{-1} for the general public. The equivalent 1990 ICRP figures given in Table 10.3 are 6 per cent and 7.5 per cent Sv^{-1} for workers and the public. These figures are somewhat higher than those evaluated by NRPB, principally because of a higher allowance for non-fatal cancers. ICRP weight the non-fatal cancer in each organ by that cancer fatality rate, leading to a contribution of 1.5 per cent Sv^{-1} for the public and 1.2 per cent Sv^{-1} for a workforce. The new ICRP organ risk factors for a population of

Table 10.4 ICRP risk and detriment factors 1990

Organ	Lethality (%)	Fatal Risk (% Sv^{-1})	Detriment (% Sv^{-1})
Bladder	0.50	0.20	0.30
Bone	0.70	0.05	0.07
Breast	0.50	0.25	0.38
Colon	0.55	0.95	1.38
Liver	0.95	0.20	0.21
Lung	0.95	0.90	0.95
Oesophagus	0.95	0.35	0.37
RBM*	0.99	0.45	0.91
Skin	0.01	0.02	0.04
Stomach	0.90	1.10	1.21
Thyroid	0.01	0.08	0.15
Remainder	0.80	0.45	0.54
Gonads	–	–	1.0
TOTAL		5.0	7.5

*Detriment figure doubled for Red Bone Marrow to account for the short latent period for leukaemia.

all ages and the associated detriment is shown in Table 10.4. ICRP is thus in a position to ensure the risk is the same whether the body is irradiated uniformly or if only partial body exposure occurs, e.g. from internally incorporated radionuclides. The quantity is called 'effectance'.

Implications for Dose Limits

The dose limits define the lower boundary of a region in which exposures are unacceptable while the requirement to keep exposures 'as low as reasonably achievable' gives a mechanism for decisions on how far below those limits it should be reasonable to strive to reduce doses. This means that it is not adequate just to demonstrate compliance with dose limits.

Given the evidence that was emerging regarding changes in the estimates of the risks of fatal cancer, NRPB in 1987 produced interim guidance for those with regulatory responsibility for setting dose limits for workers and the public, and for those involved in the management of those exposures (NRPB, 1987). It recommended that as long as the dose limit remained at 50 mSv per year the average dose to the most exposed workers should be so controlled as not to exceed an effective dose equivalent of 15 mSv per year. For the public, it recommended that, as the principal dose limit was already at 1 mSv per year, the doses to the most exposed groups from effluent discharges from nuclear installations should be so controlled as not to exceed an effective dose equivalent of 0.5 mSv per year for a single site.

It has been the practice, in radiation protection, to assume linearity between dose and risk without threshold. Therefore any exposure to radiation – however small – is assumed to carry a proportionate amount of risk. There is no reason to change this assumption: there is no convincing evidence of a response greater or less than linear. It follows that the discussion on the limitation of risk to individuals must include not only the risk per unit dose but also what risks are acceptable.

For both workers and the public, a wide range of individual risk is accepted under different circumstances. Industrial fatal risks vary world wide, but in the UK the range is from a few per million per year in shops and offices to over 100 per million per year in coal mining. The figures conceal higher risk sub-groups – quarrying gives 1 in 2,500 per year and fishermen on UK-registered deep-sea vessels have a 1 in 1,100 per year fatal accident rate. There is no single, simple dividing line to distinguish those risks that are clearly unacceptable. Rather, there is a broad range of risk within which dose limits are set for workers and the public. In order to assess what might be acceptable risks, the NRPB relied on the judgement of a Study Group of the Royal Society, which it was thought had probably reflected the current popular UK view on risk acceptability (The Royal Society, 1983). This report was not specifically concerned with nuclear matters but *all* hazardous risks. The Study Group decided that for someone of normal life expectancy, a continuing risk of 1 in 100 per year was unacceptable – you would almost certainly die from that cause. At 1 in 1,000 per year, they thought the risk may not be totally unacceptable, if the individual knows of the situation, enjoys some commensurate benefit, and everything reasonable has been done to reduce the risk. This gives an upper limit to occupational risk of about $10^{-3}\,y^{-1}$.

For members of the public, the Royal Society's judgement was that there was a widely held view that few people would commit their own resources to reduce an annual risk of death which was already as low as 1 in 100,000. If, however, there are grounds for suspecting a real risk, at an annual level of 1 in 10,000, the imposition of that risk is likely to be challenged. It seems therefore that a reflection of society's present view is that an unacceptable *imposed* risk will be one at a level between 1 in 10,000 and 1 in 100,000 per year.

The UK Health and Safety Executive (HSE) has since published its views on the philosophy of risk control in a document entitled 'The Tolerability of Risk from Nuclear Power Stations' (HSE, 1988). It is appropriate to evaluate the levels of risk corresponding to NRPB's interim guidance on restricting radiation exposure and to compare those levels with HSE's proposals.

'The Tolerability of Risk from Nuclear Power Stations' proposes an upper level to occupationally fatal risk of 10^{-3} per year. Table 10.5 shows the risk, for a population exposed at different dose-rates from age eighteen to sixty-four, and allowing for competing causes of death. The risk at age seventy and average lifetime risks can be compared with HSE's limits of tolerable risk. For workers the implied dose limit depends on how the annual risks are judged. The maximum risk now arises after retirement age

Table 10.5 Annual risks associated with levels of continued exposure

Annual dose (mSv)		Annual fatal risk	
		At age 70	Average
Occupational			
50	present limit	$3.3\ 10^{-3}$	$1.8\ 10^{-3}$
20	ICRP proposal	$1.3\ 10^{-3}$	$7.6\ 10^{-4}$
15	NRPB advice	$1.0\ 10^{-3}$	$5.7\ 10^{-4}$
Public			
5	present limit	$6.8\ 10^{-4}$	$2.8\ 10^{-4}$
1	ICRP proposal	$1.4\ 10^{-4}$	$5.7\ 10^{-5}$
0.5	NRPB advice	$7.0\ 10^{-5}$	$3.0\ 10^{-5}$

because of the use of a multiplicative model. In Figure 10.2 the time distribution of risk is shown for constant exposure at 10, 20, 30 and 50 mSv y^{-1}. The risk limit of 10^{-3} is exceeded by age fifty-five at 50 mSv y^{-1} and at age sixty-five for 20 mSv y^{-1}. A dose of 20 m Sv y^{-1} leads to an average lifetime annual risk of about 8 mSv 10^{-4}, just less than the tolerable risk limit. Assuming NRPB's interim advice for 15 mSv y^{-1} is followed, the annual risk approaches 10^{-3} at age seventy and averages 5.7 10^{-4} over the working lifetime.

For members of the public the HSE document proposes a tolerable risk limit of 10^{-4} y^{-1}. Table 10.5 shows that the present dose limit of 5 mSv y^{-1} exceeds the tolerable risk throughout the majority of life. In Figure 10.2 the time distribution of risk is shown for constant exposure over a lifetime of 1, 2, 3 or 5 mSv y^{-1}. The tolerable risk is exceeded beyond age seventy for 1 mSv y^{-1} and an annual average lifetime risk of 10^{-4} corresponds to less than 2 mSv y^{-1}. Again if NRPB advice to limit exposures to 0.5 mSv y^{-1} is followed, the risk is below HSE's suggested tolerable limit of 10^{-4} per year but probably at the upper end of the range suggested by the Royal Society's group (between 10^{-4} and 10^{-5}). At an average annual risk of 10^{-5}, the limit would be 0.2 mSv and NRPB (1987) said in GS9 that risks much beyond 10^{-5} probably verge on the unacceptable for the public. The Board's advice was for 0.5 mSv per site.

ICRP has just published its draft recommendations in which it is proposing time-averaged limits; 100 mSv in five years for occupational exposure with no more than 50 mSv in one year and 5 mSv in five years for members of the public. This implies that strictly the dose limits for workers would be 20 mSv y^{-1} and for the public, 1 mSv y^{-1}. However, from a biological point of view, a year is an arbitrary time period so that 10 mSv received one year and 30 mSv the next, is the same as 20 mSv in each year. This averaging may, however, cause administrative difficulties.

Are these limits adequate?

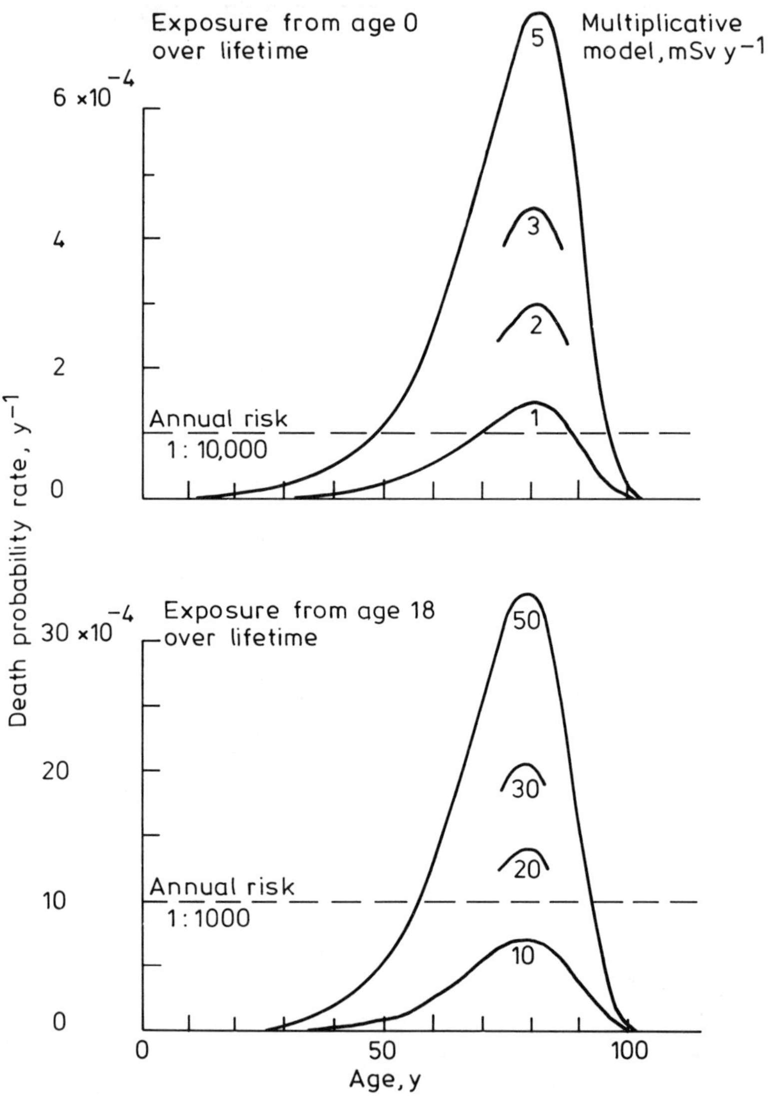

Figure 10.2 Unconditional death probability rate for two exposure situations

Note: The attributable death-age probability density is normalised for lifetime risk

Radiation Exposures in Perspective

One way of judging the acceptability of the dose limits is in comparison with the variation of natural background radiation to which we are almost inevitably exposed. The average exposure within the United Kingdom is

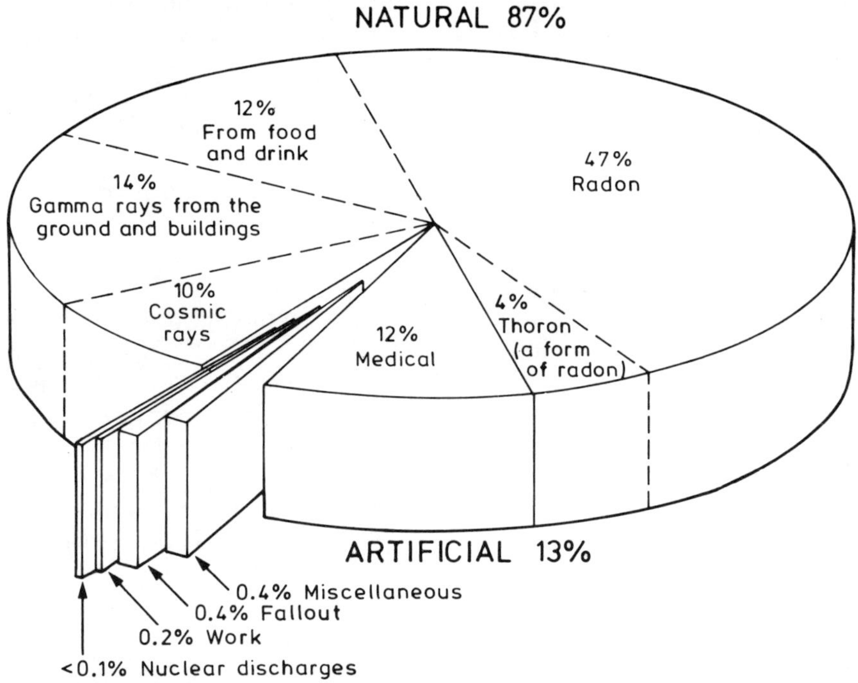

Figure 10-3 Average annual radiation dose to the UK population - 2.5 mSv overall

Source: National Radiological Protection Board, *Living with Radiation*

now 2.5 mSv/year (Hughes *et al.*, 1988). The contributions in 1988 are shown in Figure 10.3 and are, in descending order of importance, radon (47 per cent), rays from buildings and the ground (14 per cent), natural activity taken into our bodies with food and water (12 per cent) and medical exposures (also 12 per cent), cosmic rays (10 per cent), thoron gas (4 per cent), fallout (0.4 per cent), miscellaneous sources such as luminous watches and air travel (0.4 per cent), occupational exposure (0.2 per cent) and nuclear effluents (<0.1 per cent). The doses to individuals may depart markedly from the average, depending on their mode of life and geographical location. An appreciation of the extent of this variation is important in the comparative evaluation of risks.

Clarke and Southwood (1989) have contrasted, in a series of pie diagrams, doses received by different groups in the UK population, with the national average figures. The area of each pie shown in Figure 10.4 is proportional to the total dose received and the slices represent the percentages from the various sources. In general, the highest doses in the UK arise in south-west England – Cornwall and Devon. The average dose received in Cornwall is slightly over three times higher than the UK average (ie, 7.8 mSv compared with 2.5 mSv), with the inhalation of radon giving 81 per cent of the total.

There are parts of Cornwall, and some other areas of the United Kingdom, where the exposure of individuals rises to more than 20 mSv in a year when over 90 per cent of the dose is due to indoor radon. More than twenty times the number of people receive over 15 mSv y^{-1} from radon at home than receive more than 15 mSv y^{-1} at work in the nuclear industry, where the average exposure is 2 mSv y^{-1}.

The highest doses to members of the public from artificial sources of radionuclides are from the Sellafield reprocessing plant in Cumbria. Consumers of sizeable quantities of seafoods collected in the vicinity of the plant received a total exposure of 2.9 mSv in 1987, which is not greatly elevated above the national average of 2.5 mSv although doses had been higher at nearly 5 mSv y^{-1} in the early 1980s. Not all the dose above the national average is due to Sellafield however. Internal exposure is higher, due to the increased intake of natural alpha-emitting radionuclides (mainly polonium-210) which are concentrated by shellfish. This increase, which is unrelated to any man-made source of radionuclides, may raise the annual internal dose by another 0.14 mSv (from 0.30 mSv to 0.44 mSv) which is just under half the additional dose now received by those most exposed to the Sellafield discharges.

The doses to air crew illustrate one of the other features of modern life which may have a significant effect on dose received. Because of the reduction in the screening effect of the atmosphere with height, air travel increases exposure to cosmic radiation. Frequent air travellers who, in a year, fly for 100 hours (that is about five subsonic return flights between Britain and the west coast of America, or a once a week flight between London and Scotland or one of the near European countries), would increase their dose from that source from 0.25 mSv y^{-1} to 0.65 mSv y^{-1}. This would lead to a total dose, shown in Figure 10.4, of 2.9 mSv y^{-1} – the same dose as received by the most exposed member of the public from Sellafield discharges. There are many more frequent flyers than members of the critical group at Sellafield. No one is suggesting for one moment that people should not fly or be worried about the radiation exposure incurred. What is suggested is that they should be less worried about man-made sources because of the magnitude and variation of existing exposures in the United Kingdom.

Conclusions

First an assertion: people are becoming more risk conscious. Both the workforce and members of the public are able to express their concerns more numerately. They demand better standards of protection and some, though admittedly they are extremists, would like a risk-free society. This zero-risk demand cannot be met by defining radiation as having zero risk at low levels.

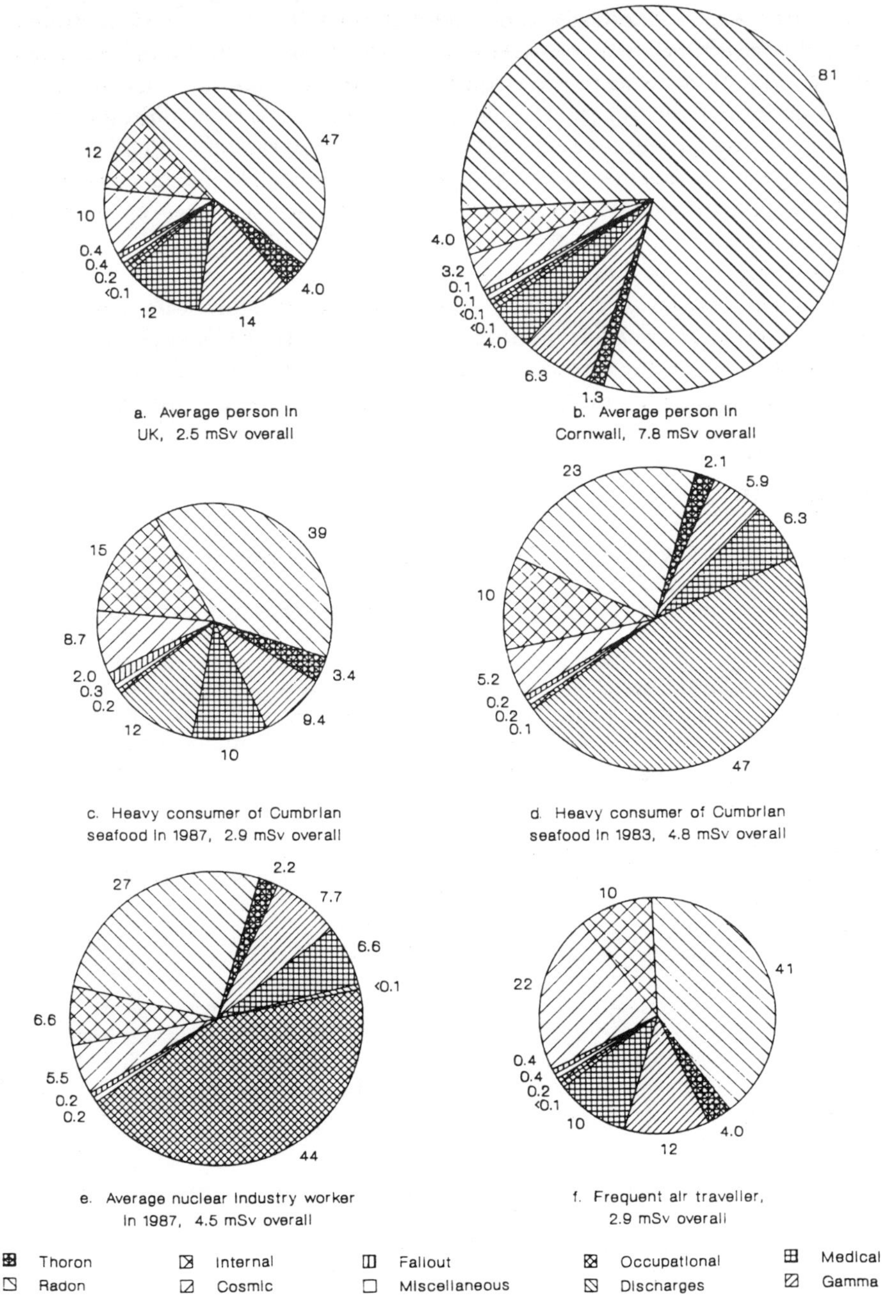

Figure 10.4 Variations in annual radiation dose in the UK compared with the average. Sizes of pie diagrams are proportional to dose and percentage segments to sources

Source: National Radiological Protection Board, Report 227.

Radiation-dose standards recommended by the Board, 15 mSv maximum annual average doses for workers and 0.5 mSv for the public in a year, are robust. Our fatal cancer risk estimates are unlikely to be underestimated. Gardner *et al.*'s (1990) recent study of leukaemia in West Cumbria and the association with paternal dose, if the result is due to radiation, reinforces the urgency for managements of radiation workforces to ensure compliance with the Board's advice. That advice can also be seen as being reasonable in the light of the variation of natural background radiation.

However, this is a time of change. Even if the risk per unit dose is thought to be fairly certain, society may seek higher safety standards that would require further restrictions on doses received. The future is uncertain, but it may be that as other environmental pollutants are assessed and controlled in the way developed for radiation, a better balance of perceptions will result.

References

British Journal of Radiology, 1934, 'International recommendations for X-ray and radium protection', 7, pp 695–9

British Journal of Radiology, 1951, 'International recommendations on radiological protection', 24, pp 46–53

Clarke, R.H., and Southwood, T.R.E., 1989, 'Risks from Ionising Radiation', *Nature*, 338, 6212, pp 197–8

Gardner, M.J., Snee, M.P., Hall, A.J., Powell, C.A., Downes, S., and Tarrell, J.D., 1990, 'Results of case control study of leukaemia and lymphoma among young people near Sellafield Nuclear Plant in West Cumbria', *British Medical Journal*, 300, 17 Feb, pp 423–9

Health and Safety Executive, 1988, *The tolerability of risk from nuclear power stations*, HSE, London

Hughes, J.S., Shaw, K.B., and O'Riordan, M.C., 1988, *The Radiation Exposure of the UK Population 1988 Preview*, NRPB-R227, HMSO, London

ICRP, 1955, 'Recommendations of the ICRP', Suppl. 6

ICRP, 1987, 'Statement from the 1987 meeting of the ICRP', *Annals of ICRP*, 17, p 4

NRPB, 1987, *Interim guidance on the implications of recent revisions of risk estimates and the ICRP 1987 Como statement*, NRPB-GS9, HMSO, London

Stather, J.W., Muirhead, C.R., Edwards, A.A., Harrison, J.D., Lloyd, D.C., and Wood, N., 1988 *Health Effects Models developed from the 1988 UNSCEAR report*, NRPB-R226, HMSO, London

The Royal Society, 1983, *Risk assessment: A Study Group Report*, The Royal Society, London

UNSCEAR, 1986, *Genetic and somatic effects of ionising radiation. 1986 report of the United Nations Scientific Committee on the Effects of Atomic Radiation*, United Nations, New York

UNSCEAR, 1988, *1988 Sources, effects and risks of ionising radiation. 1988 Report of the United Nations Scientific Committee on the Effects of Atomic Radiation*, United Nations, New York.

CHAPTER 11

Institutional Innovation to Generate the Public Acceptance of Radioactive Waste Disposal

Ray Kemp

During the 1980s, radioactive waste management came to be seen as one of the most intractable environmental issues for the majority of those industrialised countries generating a proportion of their electricity from nuclear power plants (Kemp *et al.*, 1986; Blowers and Pepper, 1987). This was so because the problem came to be recognised as being not merely a technical one, but a social and political one also. Several countries experienced intense public opposition to plans for the disposal of radioactive wastes, but this was not universally so. It has become clear that the public acceptability of radioactive waste disposal plans is a complex, multi-faceted issue, and this chapter draws upon contrasting experiences in four countries, the United Kingdom, France, Sweden and Canada, in order to throw some light on the problem of public acceptance in this field (Kemp, 1989).

In general, the difficulties encountered can be said to be a reflection of the late twentieth century conflict between science, technology and social values which affects many environment-related activities, from agricultural intensification and the extensive use of fertilisers and pesticides on the one hand, to the use of biotechnology in food manufacture, genetically manipulated organisms and food irradiation on the other. At the back of these issues is the disparity between the scientifically assessed and the publicly perceived levels of risk (Fischhoff *et al.*, 1981). The perception of risk is affected by many factors, including the level of public trust in the scientists, technologists and regulators responsible for the protection of public health and safety. It is also affected by the degree of openness surrounding a particular technology or industrial process – the accountability and consistency of decision makers, the involvement of the mass media, and the activity and commitment of environmental pressure groups (Kemp and O'Riordan, 1988). Evidence of cover-ups and political vacillation can have a detrimental effect on the public perception of the level of safety associated with a particular technology, as can the incidence and scale of accidents affecting the lives and health of vulnerable groups. But most of all, the perception of risk is at once an *individual* and a *societal* phenomenon, where public fears for the individual consequences of a risky activity quickly can become translated into concern for wider environmental and societal impacts. The so-called Not-In-My-Backyard (NIMBY) syndrome, where local opposition develops to prevent an unwarranted development planned for a particular area, can easily change into the Not-In-Anyone's-Backyard (NIABY) syndrome, where the *technology as such* is opposed on the grounds that it is environmentally unacceptable *per se*, and should not be allowed to occur *anywhere* (Kemp, 1990).

This chapter argues that a key consideration affecting the public acceptability of radioactive waste disposal is the form of the decision-making processes at work rather than the technology of disposal itself. The procedures employed in planning, negotiating and consulting with those confronted with waste-disposal plans are instrumental in affecting the public perception and acceptability of proposals at the local level. In general terms, this is true for all countries seeking to secure new radioactive waste-disposal facilities but, as we shall see, the nuances of what constitutes acceptable risk can differ considerably from country to country. In particular, this chapter highlights the fascinating differences in *institutional responses* and *innovation* designed to secure the public acceptance of radioactive waste disposal.

The United Kingdom

Radioactive wastes are classified in the UK as either high-, intermediate- or low-level wastes. Although there are certain differences from country to country, the following classification will be adhered to in this chapter. High-level wastes (HLW) or heat-generating wastes are extremely radioactive and current UK policy is for them to be stored for fifty years at Sellafield in Cumbria prior to final disposal deep underground (DoE, 1984). The wastes of immediate concern and the subject of this paper are low- and intermediate-level wastes (LLW and ILW) which are less active than HLW, and current government policy is to dispose of them as soon as possible (DoE, 1984, 1986). These wastes arise primarily as the by-products of reactor operations and fuel reprocessing. The company charged with the safe disposal of LLW and ILW is UK Nirex Ltd, though some LLW is currently disposed of in a near-surface facility at Drigg, in Cumbria, by British Nuclear Fuels.

Until May 1986, Nirex intended to dispose of LLW along with some 'short-lived' ILW in shallow, engineered trenches in one of four proposed sites identified by the company as suitable candidates for investigation. So-called 'long-lived' ILW was to be placed in a separate, unnamed site deep underground. However, following intensive local opposition at all four 'shallow' sites, and a report by the House of Commons Select Committee on the Environment in 1986 (House of Commons, 1986), the government announced that the four sites would receive LLW only. All ILW was to go to a deep-mined facility (Openshaw *et al.*, 1989). In a carefully worded statement, the Department of the Environment acknowledged that:

> . . . this is an area where it has proved particularly difficult to bridge the gap between scientists' assessment of risks and the honestly-held perceptions of the local communities.
> (DoE, 1986, p 11)

In the following year, Nirex advised the Environment Secretary that cost considerations and a concern for the need for public acceptability in its proposals, had led the company to seek a change in policy. The government agreed: all LLW and ILW for which Nirex was responsible would now be placed in a single deep repository (Radioactive Waste Management Advisory Committee, 1988).

This new chapter in the management of radioactive wastes was specifically intended to address the need for solutions which were *publicly acceptable*. Accordingly, in November 1987, Nirex launched a National Discussion Programme aimed to promote public understanding of the issues involved and to stimulate comment which would assist the development of acceptable proposals for the disposal of radioactive wastes. Between November 1987 and June 1988, some 50,000 copies of a discussion document, *The Way Forward* (Nirex, 1987) and its sister *Summary* document were distributed throughout the UK. These documents set out Nirex's proposals, outlining those facilities required by a deep repository. They presented three engineering concepts for disposal, and considered the hydrogeological environments suitable for disposal. They provided a simple base map of the UK showing the areas of geological suitability (see Figure 11.1). The proposed engineering concepts were for a facility:

> under land accessed from land; or
> under the seabed, accessed from the coast; or
> under the seabed, accessed from an offshore structure.

A range of questions upon which Nirex was seeking advice and comments were set out. *The Way Forward* stated that:

> . . . the objective is to find the right combination of site and disposal concept. A major part of the discussion process is to find out which factors are the most important in people's minds. With the benefit of these views, Nirex will be better placed to strike the right balance in selecting a site.
> (Nirex, 1987, para. 5.2.13)

Over 2,500 responses to the document were received and independently analysed by the Environmental Risk Assessment Unit at the University of East Anglia (ERAU, 1988). These included replies from over 80 per cent of county and regional councils and almost 45 per cent of district and borough councils throughout the UK. Several Members of Parliament responded, as

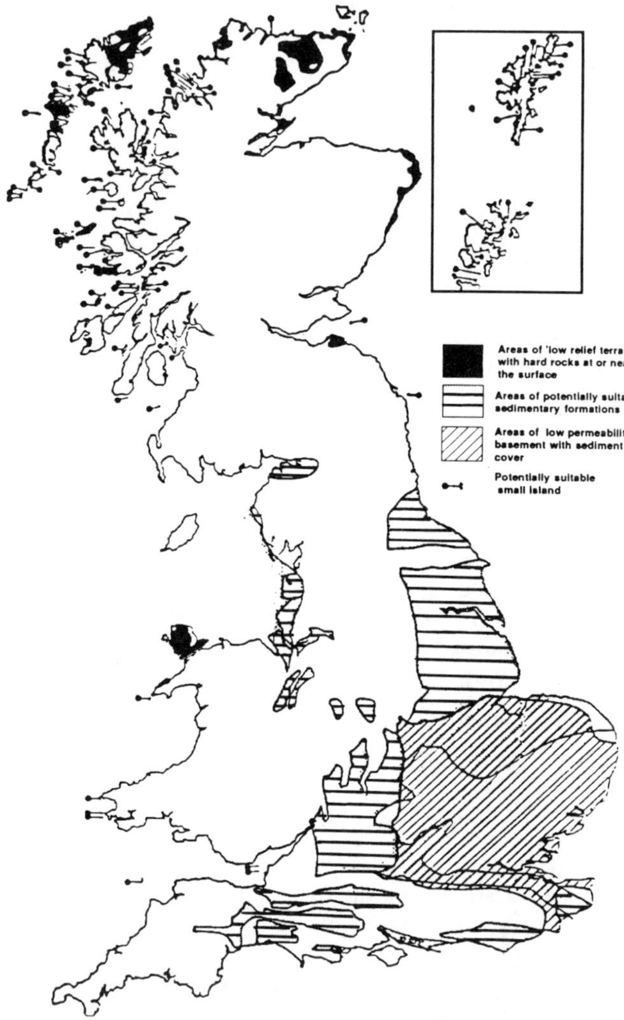

Figure 11.1 Base map showing geological environments considered to have potential for repository development

Source: Nirex, 1987, The Way Forward. A Discussion Document, United Kingdom Nirex Ltd., Didcot.

did many statutory and advisory bodies, health authorities, trade unions, and professional and scientific bodies. There was a large response from environmental groups and more than 1,700 replies came from individual members of the public (ERAU, 1988). Analysis of the responses showed that there was no overall consensus favouring any particular engineering approach or location. Environmental groups and Scottish local authorities, however, did favour an alternative approach of storing the wastes on site at their source of production. This method was not supported however by those local authorities such as Avon, Lancashire and Suffolk, which have nuclear facilities in their areas. These authorities favoured some long-term disposal solution.

The Nirex consultation exercise was generally welcomed as a major change of procedure in discussing radioactive waste management in the UK. In particular, local authorities were involved in the decision-making process at a much earlier stage than had previously been the case. It is clear that the National Discussion Programme was a unique exercise in information exchange between all the interested parties concerned.

Nirex's task was assisted in a number of key respects. First, it became clear that off-shore solutions were seen to be an unacceptable 'out of sight, out of mind' solution, and second, the discussion period raised the importance of considering the need for the possible retrievability of wastes from a deep mined facility. Retrievability of wastes was given high priority in many of the responses. Third, the exercise confirmed the importance of choosing locations with existing nuclear facilities and local knowledge of the industry, for these were the only areas in which Nirex's proposals appeared to draw a measure of public support. Fourth, public concerns were varied, but many respondents were worried about the economic consequences of perceived risks associated with the proposals (Kemp, 1990). There was some concern that a form of 'blighting' of an area chosen for radioactive waste disposal might occur. This led on to considerations of the need for local economic benefits to result from the siting of a facility to off-set any possible blight. Once again, views were mixed, with some expressing a desire for, and others a mistrust of discussions about how Nirex would seek to secure economic benefits for a community eventually hosting the radioactive waste repository.

On 21 March 1989, Nirex announced Dounreay in Caithness and Sellafield in Cumbria, both areas with existing nuclear facilities, as preferred sites for further investigation for the development of a deep mine facility for radioactive waste disposal (Nirex, 1989a) (see Figure 11.2). On 15 May 1990, Nirex successfully obtained permission to carry out exploratory drilling at Dounreay following an appeal to the Secretary of State for Scotland. A similar appeal in relation to Sellafield resulted in permission being granted for drilling work to be undertaken there in 1989. There remains local opposition to Nirex at both locations, but particularly so in Scotland. It is suggested here that future progress towards obtaining a

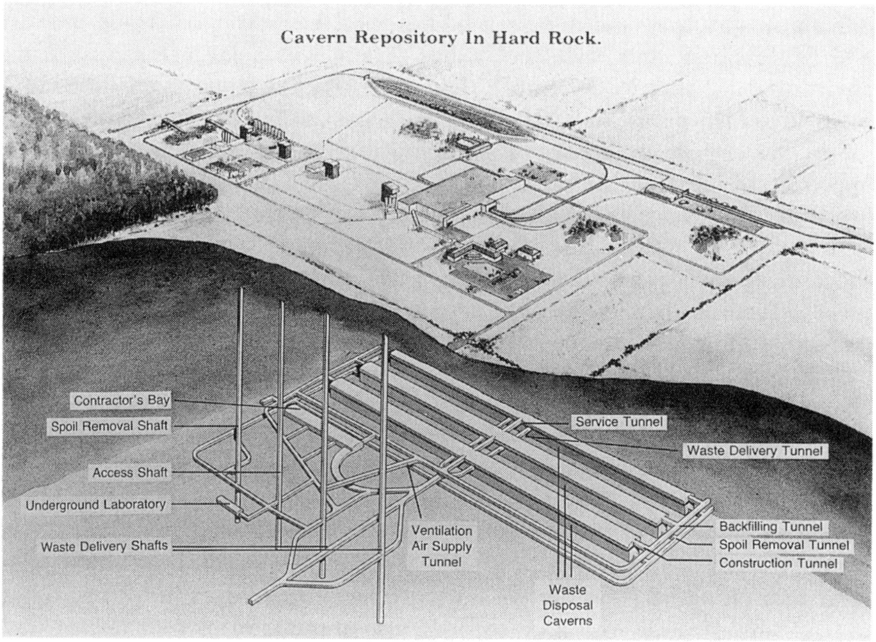

Figure 11.2 Repository beneath land – mainland access

Source: Nirex, 1989b, Deep Repository Project. Nirex Report No. 71, United Kingdom Nirex Ltd., Dicot.

measure of public acceptability for Nirex's proposals will depend to a large extent on the company's *continued* commitment to a more open and accessible decision-making process which pays careful attention to the locally perceived risks associated with radioactive waste disposal and not simply to the outcome of technical assessments. Otherwise, much of the progress gained by the National Discussion Programme will be lost and the exercise will be judged to have been a public-relations event rather than a sincere effort at obtaining public acceptance.

France

In France, ANDRA (Agence Nationale pour la Gestion des Déchets Radioactifs), which was established within the French Atomic Energy Authority (CEA) in 1979, is responsible for the long-term safe disposal of *all* categories of radioactive waste. Short-lived wastes have been disposed of by near-surface burial and above ground tumuli (in a manner not dissimilar to Nirex's original near-surface burial proposals for the UK), at Centre de la Manche, in Brittany. This site will be full by 1991 and as long ago as 1980,

ANDRA decided that a further site should be found (Kemp, 1989). In 1980, a single site at Saint-Priest-La-Prugne in the Massif Centrale was chosen. The choice appears to have been based upon confidential advice from the Bureau des Recherches Géologiques Minières (BRGM) and strongly influenced by the fact that the site was close to a uranium mine owned by COGEMA, the French nuclear reprocessing company. However, strong local opposition to the selection occurred, particularly in relation to the secretiveness of the decision-making process. M. Mitterrand championed the issue during the run-up to the 1981 Presidential election campaign, and the site was dropped in May 1980.

The next stage in the search for a near-surface disposal facility began with the establishment of the Castaign Commission in 1983. This Commission was intended to provide much broader, independent, technical advice. Its findings clarified the geological criteria to be applied for the new site; argued for an eventual choice to be made between several sites; and suggested that two future sites, not one, should be established (Castaign, 1984). Accordingly, between 1984 and 1985, ANDRA began a further search for suitable sites, investigating three potential geological 'zones'. ANDRA's new search was now a much more public process and indeed one additional 'self-candidate' zone was investigated, where local representatives had approached ANDRA, inviting the company to consider its suitability for a waste disposal site.

Moreover, as ANDRA refined its search within these zones down to actual potential sites, two self-candidate sites were investigated – at Neuvy-le-Roi and at Cholet. ANDRA's final choice was, however, at Soulaines-Dhuys in the Aube region, to the east of Paris, and plans were submitted to government for approval in 1986. Late in the year, a local Enquête Publique was held, essentially an information exchange and gathering exercise for the Environment Ministry, and permission to develop the Soulaines site was granted to ANDRA in the summer of 1987. The new surface facility will become operational in 1991 (see Figure 11.3).

The factors which appear to have been instrumental in guaranteeing a measure of public acceptability for ANDRA's proposals to develop a new near-surface radioactive waste-disposal facility are clearly a mixture of features peculiar to the French situation and ones which have more universal applicability. First, the French general public's perception of nuclear power is patently different to that in most other countries. With over 70 per cent of French electricity production generated from nuclear reactors, the position of nuclear energy within the national political framework is unique. Moreover, France has had extensive experience of the near-surface disposal of radioactive waste at Centre de la Manche. The establishment of commissions d'information, or community liaison groups, around all French nuclear sites has also facilitated the exchange of information about both the risks and benefits associated with nuclear

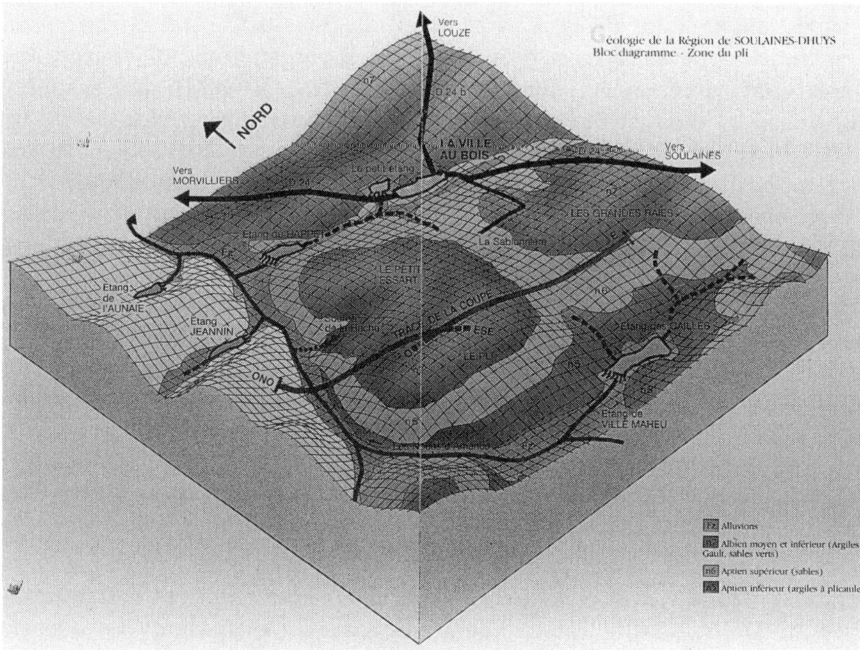

Figure 11.3 Near-surface disposal: the necessary local geology for the chosen site at Soulaines (top); disposal at Centre de la Manche (bottom).

Source: ANDRA 1981 *Centre de La Manche* Agence Nationale Pour La Gestian des Déchets Radioactifs, Paris: ANDRA 1986 *Site de L'Aube* Agence Nationale Pour La Gestian des Déchets Radioactifs, Paris

operations among the general public. Indeed, ANDRA encouraged contacts between the Centre de la Manche commission and the local communities at Soulaines during the new-site selection process.

Second, and leading on from this, ANDRA developed extensive information campaigns to publicise the benefits which would accrue to any host community. Payments by firms into local community budgets are common in France, much more so than in the UK, where it is comparatively unknown. The three small communities close to the new site in the Aube region will receive a capital payment of 30 million francs (1984 prices) plus 1 million francs per annum. These monies are to be divided as follows:

33 per cent to three local villages;
33 per cent to the local region;
33 per cent for communal programmes.

These are seen to be the 'natural' economic benefits which should accrue from the development to help with the local infrastructure – the provision of better medical services and school buildings for example, which will be necessary to support the increasing population consequent upon the construction and operation of the site.

Thus it can be seen that ANDRA's experience in the early 1980s confirmed the importance of open decision-making/site-selection processes and of addressing the issue of the local economic impact of the proposed facility. While the 'self-candidate' phenomenon may appear rather strange to the Anglo-Saxon reader, the basis of ANDRA's approach was clear – a strong commitment to the technical programme, to creating good conditions for local communities, and a willingness to respond to the criticisms of earlier actions with an innovatory, independent, institutional 'fix' – the Castaign Commission, which set the ground rules for ANDRA's later success. As we shall see below, this combination of independent review and the encouragement of volunteer sites occurs with much greater emphasis elsewhere, as a possible means of generating the public acceptance of disposal plans.

Sweden

The situation in Sweden provides an interesting comparison with both the UK and the French nuclear power and radioactive waste management systems. With a full complement of twelve reactors, Sweden's nuclear-power programme is closer in size to the UK than to France. But in terms of overall dependency on nuclear energy as a fuel source, Sweden more closely resembles the situation in France, with a little over 53 per cent of Swedish electricity production coming from nuclear stations in 1985 (Parker *et al*,, 1987; Kemp, 1989).

The political context of nuclear-energy policy in Sweden is also different. Following the 1979 Three Mile Island accident in the United States, a national referendum in Sweden in 1980 narrowly resulted in the adoption of a policy to end nuclear-power generation there not later than the year 2010. As a consequence, the size of the nuclear programme and its waste arisings are well understood, and the Swedish nuclear-power industry has made better progress than most other countries in establishing radioactive waste-disposal facilities.

Chief among these is the Central Final Storage Facility (SFR) for LLW and ILW in a rock cavern at Forsmark. This fully operational facility is a series of mined caverns and an engineered, 70 m high, concrete 'silo' some 60 m below the sea bed. The SFR is accessed via two 1 km long tunnels from the coastal harbour facilities close to the site of the three existing Forsmark nuclear reactors (Figure 11.4).

The Swedish Nuclear Fuel and Waste Management Company, SKB, confined its search for a suitable location to the four Swedish nuclear power station sites (Oskarshamn, Barsëback, Ringhals and Forsmark) on the basis that it had good relations with the local communities which appreciated the employment opportunities and economic benefits associated with the industry. In addition, there were good handling facilities, a workforce with radiation protection experience and a nuclear-related industrial and commercial infrastructure. The sites were also in areas with a suitable geology, but the east coast sites were preferred for having less fractured rock. Eventually, Forsmark was chosen, because the third reactor construction programme was coming to an end and a suitable workforce was therefore available to begin construction on the SFR in 1987. The SFR construction ended in 1988 and a site licence was obtained in 1989.

Within the international nuclear community, the SFR at Forsmark is considered to be one of the few real successes of radioactive waste management. To support a comparatively small reactor programme, the SFR concept may be said to have been ambitious and relatively expensive. It is a deep site, in hard rock (granite), with a combination of natural and engineered barriers designed to contain radioactivity within the site. Taking some 90,000 m^3 of LLW and ILW, the SFR is approximately one-tenth of the size of the deep repository sought by Nirex in the UK. But the SFR is operational, and SKB are currently in the process of identifying a suitable site for the final storage of HLW – far in advance of many of their counterparts.

The factors which appear to have been influential in obtaining a measure of public acceptability for Sweden's waste-disposal programme are once again a combination of technical and political considerations. It is undoubtedly true that the 1980 referendum created a special context for nuclear power and waste-management decisions in Sweden. Great emphasis was placed on the need for safety and 'over-designing' in the waste-management programme. Arguably, this has increased the costs but reduced

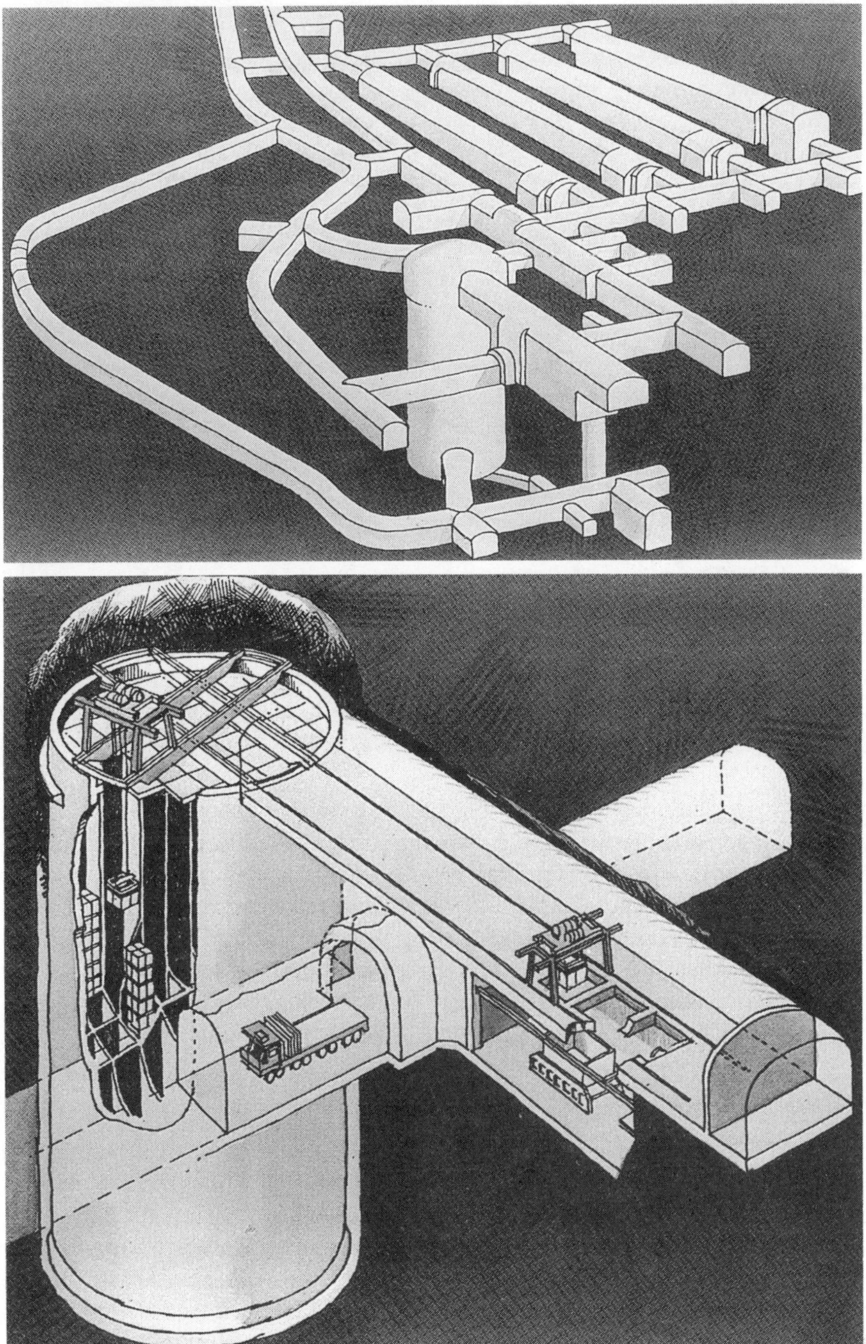

Figure 11.4 The Central Final Storage Facility at Forsmark

Source: SKB (1988) *Slutförvar För Radioktivtdviftavfall-SFR* Svensk Kärnbränsle–Hautering, Stockholm

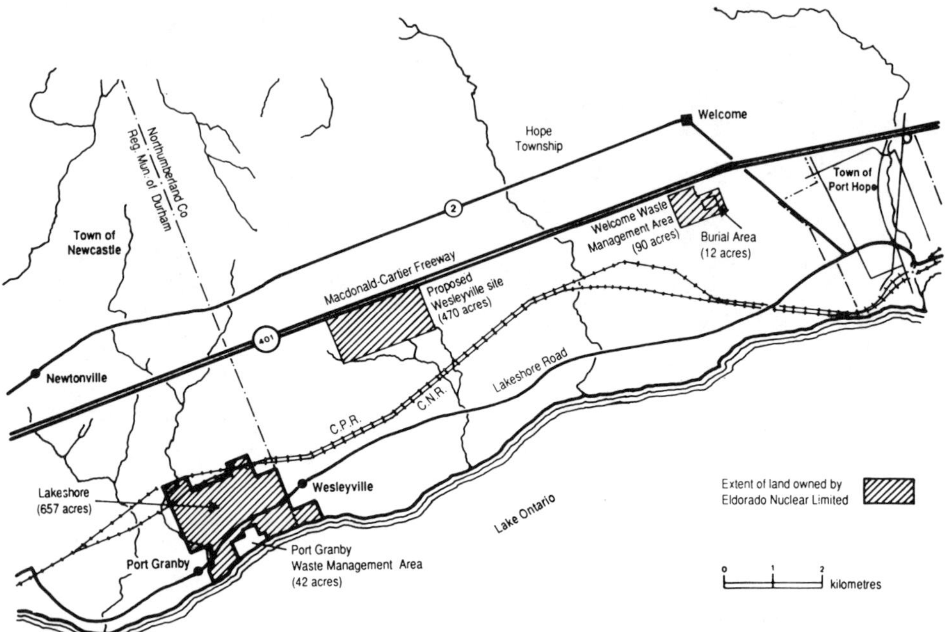

Figure 11.5 Port Hope area low-level radioactive waste sites

Source: Siting Process Task Force, 1987, *Opting for Co-operation. Report of the Siting Process Task Force on Low-level Radioactive Waste Disposal*, Energy Mines and Resources, Ottawa, Canada

delays in the construction and licensing programmes. Second, site selection was confined to those areas with existing nuclear facilities and therefore to those areas with a level of familiarity with the nuclear industry and where a level of local economic dependency on the industry had developed. Should Sweden need to review its overall energy policy priorities again as the year 2010 approaches, the Swedish nuclear-power industry will have resolved many of the problems of the 'back end' of the nuclear fuel cycle which create difficulties of public acceptability in so many other countries.

Canada

The final example of attempts to obtain public acceptance for a radioactive waste-disposal proposal is from Ontario, Canada. The problem here is slightly different in that it does not concern wastes arising directly from nuclear-power operations, but rather historic wastes from uranium-mining activities in the Welcome, Port Hope and Port Granby areas on the northern shore of Lake Ontario (see Figure 11.5). Here, some 600,000 m^3 of historic LLW and contaminated soils need to be disposed of in a new facility because of problems of long-term bluff erosion along the site of the existing facilities which are managed by Eldorado Resources Ltd (ERL) (Siting Process Task Force, 1987).

In 1982 ERL, having been directed by the Canadian Atomic Energy Control Board (AECB) to develop a decommissioning plan for the Port Granby and Welcome waste areas, announced a proposal to establish a new disposal facility for the wastes using either engineered burial or disposal caverns, not dissimilar to the types of concept considered during the 1980s by UK Nirex Ltd. ERL initiated consultations with committees representing the two municipalities of Port Hope and Port Granby in 1982. ERL's experience in consultation with the local communities was not a happy one. One committee voted to break off discussions leaving the second committee feeling that it was being presented with a *fait accompli*, so it too withdrew from the discussions. In 1984 ERL went on to announce five potential sites and obtained AECB approval for its design concepts.

However, federal elections in 1984 led to renewed political interest in the issue and subsequently a government directive on 25 April 1986 forced ERL to choose potential sites no nearer than 5 km to Lake Ontario, but to the north of the area's main highway. In due course, ERL announced Newcastle North as its preferred site and initiated a $3 million investment programme. Strong local opposition to the ERL choice developed. Petitions were sent to the federal government and several effective local protest demonstrations took place. Later in the year, ERL's plans were rejected by the federal government and, in a major innovation, responsibility for site selection was placed in the hands of a new Ministerial Task Force in December 1986.

The Siting Process Task Force was to be an authoritative body made up of independent experts with a mandate to 'design a less confrontational process for the siting of a low-level radioactive waste facility' (Siting Process Task Force, 1987). In particular, the Task Force was asked to develop a *co-operative* process for facility siting with the main objective being the *voluntary* identification of a possible host community and associated disposal technology. This was a significant procedural innovation with the express purpose of obtaining a solution to the radioactive waste disposal problem which was not only technically appropriate but at the same time locally acceptable.

In June 1987, the Task Force produced a five-phase Draft Consultative Site Selection Process which was based upon six key principles (Figure 11.6). These were first, that a community should volunteer rather than be selected by the project developer and should retain the right to opt out of the process at any stage. Second, participating communities should do so on the basis of equal partnership in decision making. Third, that communities should be compensated for 'unmitigable impacts' associated with the development and should receive local community benefits for assuming the burden of any latent risks associated with the facility. Fourth, that the communities should be able to select the technical disposal option to be employed and fifth, that the Task Force should ensure that there would be no compromise on environmental and human safety criteria. Finally, that the 'soft issues' of public perception of the risks associated with the facility,

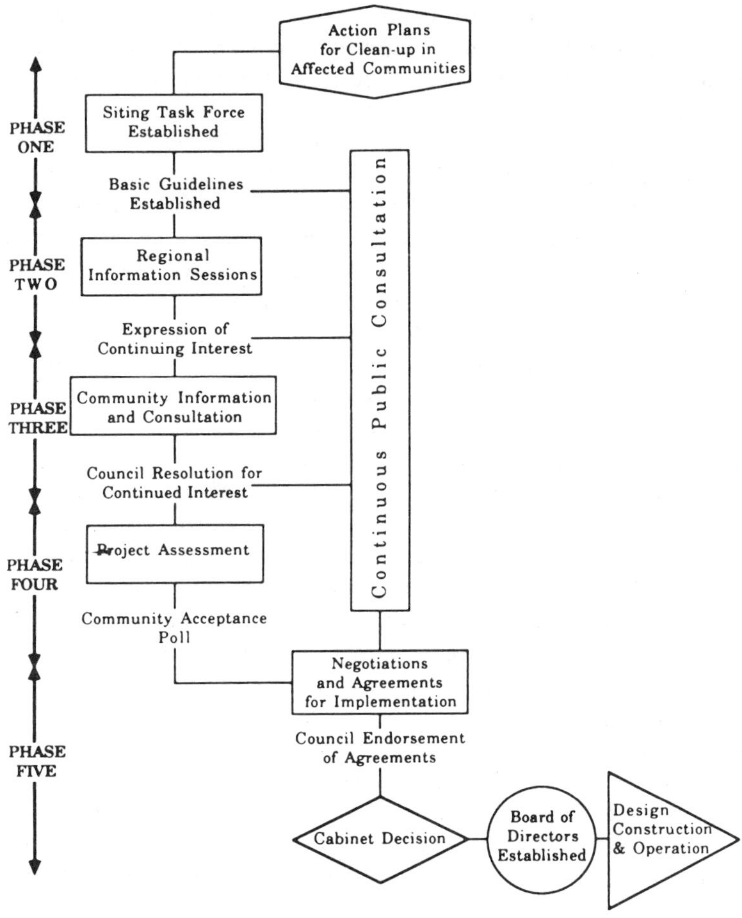

Figure 11.6 A consultative siting process

Source: Siting Process Task Force, 1987 – as for Figure 11.5.

equity issues and planning blight, the need for local control over the facility and establishment of trust in both the operators and the regulators of the facility, also needed to be addressed during the site-selection process.

In summary, the philosophy behind the approach was that, once safety assurance was achieved, community interests and concerns should be dealt with as actively as technical issues. The Task Force's proposed five-phase process utilises independent, expert peer review; local public meetings for information dissemination and consultation; local community liaison groups for discussion of technical options; joint fact finding and problem solving using cost/benefit analysis, environmental impact assessment and related techniques; and community acceptance polls and resolutions prior to any final decision being taken by federal government at Cabinet level. Finally, the facility, once operational, would be overseen by a Board of Directors who would include local community representatives.

In September 1988, these radical proposals were accepted by the Canadian Cabinet and a new Task Force was established to implement the siting process. Initially, fourteen communities voluntarily expressed an interest in participating in the programme. By 1990, ten communities remained in the process with Phase 3 of the overall programme coming to a close in May 1990. Cabinet permission will be sought prior to the start of Phase 4 of the process – Project Assessment.

Ironically, of those communities remaining in the new site-selection process in 1990, Hope Township, Newcastle and Port Hope have been active in the discussions, with Port Hope appearing to be the most likely eventual host for the facility, having originally broken off discussions with ERL in 1982. Clearly, the innovatory approach adopted in Canada has been costly both in terms of time and resources committed to the exercise. Nevertheless, it has met with some success in maintaining programme momentum while at the same time creating a greater sense of public involvement and openness. Of all the site-selection procedures for radioactive waste facilities explored in this paper, the Canadian process displays the most innovatory features designed to address the question of the public acceptance of both the decision-making process *and* the waste-management technology. In the next section, we shall consider those elements which are common to all the different examples covered and offer an interpretation of the key factors which impinge upon the question of the public acceptance of radioactive waste-disposal plans.

Conclusions

The public acceptance of innovation, in this case the development of new facilities for the disposal of radioactive wastes, is a complex, multi-faceted phenomenon. It is mistaken to assume that technical competency of the very highest standards and scientific assessments of the risks associated with a particular operation or development will, or should even be expected to, generate public acceptance. Nor is it true to assume that public perceptions of risks are entirely subjective, context-driven and socio-cultural phenomena which can only be assuaged by political processes (Wynne, 1983). Rather, our current understanding of the nature of risk, of the relationship between technological assessments and societal judgements, is that the two aspects are inextricably interwoven (Brown, 1989). If decisions are to be implemented then the establishment of public confidence in both the

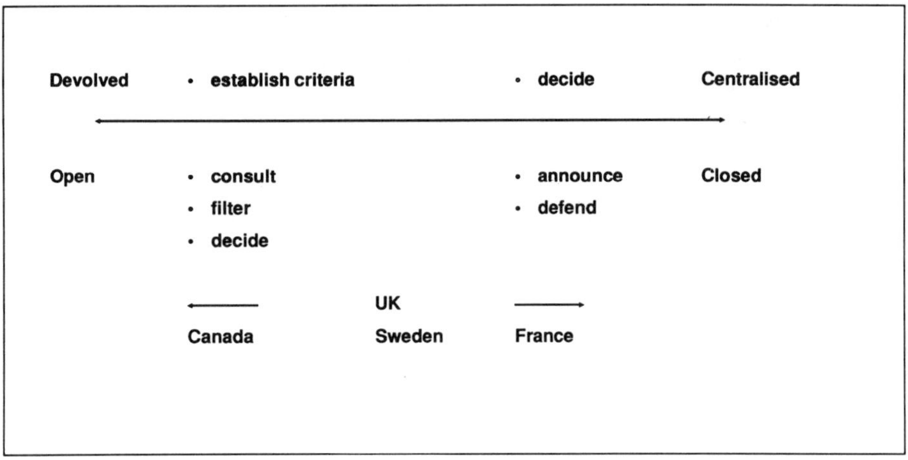

Figure 11.7 A spectrum of procedures

practices of the industry in question and the planning and consultation processes at work is essential. In other words, acceptable risk is not merely a function of sound science and good engineering practice. It depends on a combination of risk assessment, risk regulation and management, and risk communication where the principles of comprehensibility, openness, accessibility and equity are paramount (Kemp, 1991; Kemp and O'Riordan, 1988).

A central conclusion of this comparative review is that although the emphasis on these key principles will differ from one national context to another, their relevance to public acceptability remains unquestioned. Thus, for example, the emphasis placed upon the role of 'honest brokers', access to technical documentation and compensation for residual risks will differ from country to country, but aspects of each will be present throughout.

That is not to say that there is some universal panacea or process guaranteed to elicit public acceptance. It is clear that institutional adaptation and innovation designed to address the problem of the public acceptability of risk-related proposals can have a measure of success. And in part that success will depend upon the constancy with which a particular approach to the public is maintained. Thus, for example, Figure 11.7 presents a spectrum of procedures designed to generate the public acceptance of radioactive waste disposal. The dichotomy here is between the 'devolved' and 'open' as opposed to more 'centralised' and 'closed' decision-making procedures. The important factor then, is not the position of a particular country along the spectrum, but the *direction* of its procedures and the consistency with which that direction is maintained. Once decision makers appear to lack consistency and policies appear to change for unclear motives, once technical assessments of risk are revised or embroiled in scientific dispute, then public perception of the risk can become intransigent. Once the social amplification of risk, as Kasperson and his colleagues have termed it (Kasperson *et al.*, 1988), takes hold, it becomes very hard to redress the balance. Hence the extensive efforts in both Canada and the UK to develop novel institutional procedures to address more directly the problem of public acceptability.

As this chapter has attempted to demonstrate, the practical difficulties of designing innovatory institutional procedures to address the problem of public acceptability are not inconsiderable. Such processes can be as complicated and as crucial to the final outcome as the technology assessments and scientific processes which give rise to public concerns in the first place. The final test of public acceptability, of course, relates to the question of the *legitimacy* of the decision-making process itself, i.e. the extent that decisions relating to processes which entail some level of societal risk, such as radioactive waste disposal, are arrived at in a truly open, accountable and legitimate manner. If this is so, then there is every expectation that at least in the longer term, public acceptance may be achievable.

References

Blowers, A., and Pepper, D., 1987, *Nuclear Power in Crisis*, Croom Helm, London

Brown, J. (ed.), 1989, *Environmental Threats*, Belhaven Press, London

Castaign, M., 1984, *Rapport du groupe de travail sur les Recherches et Développements en Matière de Gestion des Déchets Radioactifs*, Ministère du redeploiement industriel et du commerce extérieur, Paris

DoE, 1984, *Disposal Facilities on Land for Low and Intermediate Level Radioactive Wastes: Principles for the Protection of the Human Environment*, HMSO, London

DoE, 1986, *Radioactive Waste. The Government's Reponse to the Environment Committee's Report*, HMSO, London

House of Commons Select Committee on the Environment, 1986, *Radioactive Waste*, HMSO Cmnd. 253, 3 vols, London

ERAU, 1988, *Responses to 'The Way Forward*', Environmental Risk Assessment Unit, University of East Anglia, Norwich

Fischhoff, B., Lichtenstein, S., Slovic, P., Derby, S.L., and Keeney, R.L., 1981, *Acceptable Risk,* Cambridge University Press, Cambridge

Kasperson, R.E., Renn, O., Slovic, P., Brown, H.S., Emel, J., Goble, R., Kasperson, J.X., and Natick, S., 1988, 'The Social Amplification of Risk: a conceptual framework', *Risk Analysis*, 8, 2, pp 177–87

Kemp, R., 1989, *Planning and consultation procedures for low-level radioactive waste disposal: a comparative analysis of overseas experiences*, Research Report No. 1, Environmental Risk Assessment Unit, University of East Anglia, Norwich

Kemp, R., 1991, 'Risk tolerance and safety management', *Reliability Engineering and System Safety* 31, 3, pp 345–53

Kemp, R., 1990, 'Why Not In My Backyard? A radical interpretation of public opposition to the deep disposal of radioactive waste', *Environment and Planning, A*, 22, pp 1239–58

Kemp, R., and O'Riordan, T., 1988, 'Planning for radioactive waste disposal. Some central considerations', *Land Use Policy*, January, pp 37–44

Kemp, R., O'Riordan, T., and Purdue, M., 1986, 'Environmental politics in the 1980s: the public examination of radioactive waste disposal', *Policy and Politics*, 14, 1, pp 9–25

Nirex, 1987, *The Way Forward. A Discussion Document*, United Kingdom Nirex Ltd, Didcot

Nirex, 1989a, *Going Forward.* United Kingdom Nirex Ltd, Didcot

Nirex, 1989b, *Deep Repository Project*, Nirex Report No. 71, United Kingdom Nirex Ltd., Didcot

Openshaw, S., Carver, S., and Fernie, J., 1989, *Britain's Nuclear Waste. Safety and Siting*, Belhaven Press, London

Parker, F.L., Kasperson, R., Anderson, T-L., and Parker, S.A., 1987, *A Comparative Study of Selected Radioactive Waste Policy Issues: technical and sociopolitical considerations*, The Beijer Institute, Stockholm

Radioactive Waste Management Advisory Committee, 1988, *Ninth Annual Report*, HMSO, London

Siting Process Task Force, 1987, *Opting for Co-operation. Report of the Siting Process Task Force on Low-level Radioactive Waste Disposal*, Energy, Mines and Resources, Ottawa, Canada.

Wynne, B., 1983, 'Public perceptions of risk: interpreting the "objective versus perceived risk" dichotomy', International Institute for Applied Systems Analysis, Working Paper 83/117, Laxenburg

CHAPTER 12

Variation in Information Use in the Evaluation of the Social Acceptability of Nuclear Waste Management Options

P. T. Allen

This chapter describes an attempt to provide one answer to the following question: what use do people make of information, especially technical information, when they are attempting to form judgements about complex technological phenomena? The question is highly relevant to the issue of the public understanding and acceptance of innovations, the more so where these have implications for health and safety.

Within many modern institutions there are formal and semi-formal procedures for choosing courses of action which often include prescribed methods for the evaluation of information. This contrasts strongly with the normal methods of evaluation by the public which are typically not governed by formal rules and which are usually based on only partial information.

Attempts to incorporate or take account of the public response frequently focus on this lack of information and there are calls for more effective ways of providing technical information and 'educating the public'. The notion of education also implies a need to take account of the ability of the public to *use* technical information in appropriate ways, since it is clear that, in general, people will lack more than the information specific to particular technological strategies. Thus, there is concern that wider scientific education has been lacking and that a proper understanding of the scientific perspective would ameliorate the difficulties posed by an apparently uncomprehending public.

However, there are at least three reasons why this straightforward approach may be ill conceived. Firstly, many of the situations, processes or developments which cause concern involve political, social, moral or ethical aspects which are clearly not matters which can be resolved by recourse to technical data. With reference to these aspects any person in a modern democracy who might be judged legally competent must be accorded equal status as a judge of acceptable practice.

Secondly, in any situation in which there are political, moral or ethical problems there are also likely to be vested interests. People with such interests will select and edit information in order to maximise some desired end such as a general aim, or simply to enhance public support or to gain social or political power.

Thirdly, and perhaps less obviously, it is not at all clear that technical information can *ever* be reliably used, in an informal evaluation, in the ways which are sometimes characterised as 'rational' and hence acceptable, or at least defensible, when incorporated into institutional policy. Indeed, it is part of the reason for the development of decision theories, and their rule-governed or *normative* procedures, since *any* less formal evaluations will be sub-optimal.

A considerable tradition of research has built up which has identified many of the features of what Simon (1955) called the 'bounded rationality' of human beings. Field-based research has identified inadequate information use in, for example, policy formation (Lindblom, 1964) or business (Slovic, 1969), and theoretical and experimental research has begun the task of mapping the discrepancies between axiom-based strategies (von Neumann and Morgenstern, 1947; Raiffa, 1968; Keeney and Raiffa, 1976) and actual human techniques and performance (Edwards, 1965; Tversky and Kahneman, 1974).

This chapter focuses on the use of information by people judging the social acceptability of nuclear waste management options. It reports findings from a field-based experiment designed to illustrate the difference between 'normal' and 'normative' evaluation procedures used by both technically experienced and lay participants.

The Context of the Research

The experiment which forms the core of this chapter was part of a much larger study of social values in the context of nuclear waste management. In common with many modern technological enterprises, but perhaps to an even greater degree, the nuclear industry is faced with the need to take account of non-technical evaluations in the management of its activities.

In order to take a step forward in this uncertain arena, attempts are being made to develop structured approaches which may incorporate input from the public (Lathrop and Watson, 1982). Thus, the concern of the National Radiological Protection Board, for example, is to go beyond the setting of dose limits to the development of methods for establishing criteria which might enable optimal choices to be made in the minimisation of radiation exposures *and* economic *and* other social costs (Hart, 1988).

One promising direction which might help achieve such aims involves the use of multi-attribute utility analysis (Edwards, 1977; von Winterfeldt and Edwards, 1986). Within this family of techniques there are various levels of sophistication. But, in general, since the elements of any decision relating to radiological impact are not measured in units commensurable with social values, or aggregated estimates of these, the expert technical assessments have to be evaluated along parallel subjective dimensions. One of the psychological tasks was therefore to devise ways to scale these evaluations.

In one very simple format, a multi-attribute analysis could be conducted by deriving multipliers, or weights, for each relevant technical measurement, from a range of public views, and then combining these with the technical estimates using a linear combination rule. Some options might prove to be invariant despite differing social values on some attributes, such as cost. Alternatively, sensitivity analyses using different values in the linear equation might reveal particularly important attributes which could receive additional attention.

To establish the importance of each attribute, to give it its weight, was thus a major task of the research. In the context of radioactive waste, the weight of an attribute must be judged in terms of its contribution to 'social acceptability'. This concept was therefore a focus for the participants in the research.

The perspective which was adopted for part of the project was that of social judgement theory. Its relevance, both to the research itself and to the debates about public response, arises because it seeks to make plain the ways in which people select and use information in a social policy context. Hammond and Adelman (1986) have argued that the analytical combination of attribute evaluations is socially responsible because it is an overt process, and avoids the evaluations being partly determined by existing option preferences.

Thus the multi-attribute analysis not only improves on the mechanisms of informal evaluation by increasing the reliability of information use, at least at the decision stage, but also provides a visible and socially just procedure. The technique avoids prejudice and hence fits with Rawls' (1971) insistence that attributes should be evaluated generally, and not in terms of how well the anticipated results of separate evaluations might affect the preference order. In other words, people cannot (or should not be able to) choose to evaluate the separate components of some course of proposed action in such a way as to ensure that their pre-existing preferred option comes out best. This is similar in principle to those rules of evidence which prohibit the availability of certain information likely to prejudice judgements.

However, people *do* use information in just such ways and social judgement theory recognises this, based as it is on the Brunswikian tradition which argues for the validity of psychological approaches in which human judgement is embedded in the environment (Brunswik, 1956), so that people make sense of phenomena in the context of the phenomena.

Work in this tradition has therefore focused on the ways in which the individual's environment affects judgement. But the main techniques of this form of analysis, the use of multiple correlation between environmental 'cues' and individual judgements of that environment, have also been used to examine or model the individual's policy across a set of relevant attributes. In this way the linear combination of attribute values, which is used as a normative rule in a multi-attribute analysis, becomes a model of individual judgement policy in social judgement theory. The essential difference is that in the social judgement paradigm the weights attached to attributes are inferred from a statistical analysis of the responses to the judgement task presented to the individual. Such tasks involve judgements about a set of options or scenarios which are defined by explicitly described atttributes, usually in metric form. In the social policy context such attributes may be quantified by expert technical assessment and hence the aggregated values of interest parties may be subsequently combined with the technical estimates.

By deriving weights from analyses of responses to relatively realistic scenarios judged in the round, the social judgement approach permits the individual to respond in a more natural manner than might be the case if weights were requested in a direct way, and separate from any particular context. The modelling of judgement using overall evaluations of options may therefore capture more of the influence of the wide social environment which informs the value structure of the individual.

Indeed, many studies in the UK have confirmed the existence of differing but coherent belief patterns in the fields of nuclear energy and waste management which can be seen as part of wider patterns of response to social issues (Lee *et al.*, 1984; van der Plight, Eiser and Spears, 1986; Lee, 1987; Brown and White, 1987).

The basic procedure for the modelling of social judgement is as follows. Respondents are first given some general information on the social policy problem. In this case people were supplied with a summary about radioactivity and related facts about collective doses and waste management. People are then given information on a number of options, expressed as impact estimates across several attributes, such as cost. The respondent is asked to read the information and to award a score to each option on some arbitrary scale of social acceptability.

The resulting scores are then treated as a sample of that individual's policy with respect to the use of the information supplied on the attributes. In practice a statistical regression analysis is performed using the option evaluations as a dependent variable, and the information on the attributes as independent variables. The result of this stage is a linear equation for each person which gives a set of coefficients representing the relative importance of each attribute, in terms of its contribution to explaining the variation in the individual's option scores.

In addition, the equation can supply a set of predicted option scores, which, in most cases, will not exactly match the actual scores awarded by individuals. The correlation between these predicted scores and the actual scores is known as the multiple correlation coefficient (R), which in this case is an index of the fit between the linear model and the actual scoring procedure of the respondent. In fact the square of this coefficient, known as the 'explained variance', is directly interpretable as a measure of the predictability of the individual response, and hence of the use made by the individual of the information supplied on the attributes (Hammond and Summers, 1972).

Public and Experts; An Experimental Design

Once the individual regression analyses have been conducted each respondent may be allocated the value of R^2 as a score representing the predictability of his or her policy. It thus becomes possible to examine these values in terms of the characteristics of a large number of respondents.

In order to illuminate the similarities and differences between people judging such complex proposals, an experiment was designed to quantify the effects which might exemplify the variation between 'normal' judgement, and a more nearly 'normative' procedure. In addition, respondents were chosen both from the nuclear industry and from the public so that the effects of experience with the technology might be included as a further aspect of the experiment.

The two factors chosen to provide variation in the responses reflected the main distinctions between formal and informal evaluation. The first factor, Labelling, concerned the nature of the provided information on attributes. In one condition people were just given the technical estimates of the option impacts. This information was provided by the NRPB and, where necessary, was translated from scientific notation into a form which made fewer assumptions about the experience of respondents (Barraclough and Mobbs, 1988). For example, dose rates were expressed as fractions of the current UK annual average natural background radiation, in order to avoid the use of Sievert units.

In the other condition for this factor, respondents were given the same technical estimates but, in addition, were given names or labels for the options, together with brief descriptions of the techniques involved. Twelve such options were presented, six each for intermediate- and high-level radioactive waste (ILW & HLW). These were:

Sea disposal (ILW)
Deep inland burial (ILW)
Shallow land burial (ILW)
Coastal tunnel disposal (ILW)
Off-shore borehole (ILW)
Long-term storage followed by deep inland burial (ILW)
Coastal tunnel (HLW)
Deep inland burial (HLW)
Off-shore borehole (HLW)
Ship-released ocean-floor penetrators (HLW)
Drilled emplacement in the ocean floor (HLW)

Long-term storage followed by deep inland burial (HLW).

The brief descriptions which accompanied these labels were designed to be as neutral as possible and referred to the types of technology involved. The options were chosen to represent a variety of impacts across several attributes. There were seven of these:

1 Geographical spread, and hence size of population receiving a collective dose
2 Cost in £m
3 Probable deaths from occupational exposure
4 Maximum annual risk of death to individual members of the public
5 The average dose-rate for the first thousand years
6 The average dose-rate for between one and ten thousand years
7 The average dose rate after ten thousand years.

The second factor concerned the effects that might be expected when respondents are either encouraged to consider each attribute separately before evaluating the options, as in a formal evaluation procedure, or are merely asked to evaluate options without any prior practice, as might be the case in a simple public debate. This manipulation therefore involved the use of an additional procedure which was called the Dimensional Rating Task (DRT).

The DRT consisted of a set of instructions, a scoring sheet, twenty-one blue plastic magnetic rectangles and a steel clipboard. The scoring sheet was printed with seven columns, one each for the attributes that had been selected to characterise the options. Each column was numbered from zero to twenty-one and the instructions guided respondents to build their own histogram or bar chart on the sheet, with the height of each column representing the importance that the respondent thought appropriate for each attribute in the context of a decision on nuclear waste management. This task served a dual purpose in the research since it also provided for a strictly relative judgement about the weights of the attributes for possible use in a multi-attribute analysis. In the experimental design its value was that it provided an opportunity for some respondents to practise separate attribute evaluation before judging whole options. Thus one half of the respondents were given the DRT before they came to evaluate options, and the other half were given it after their option evaluations.

These two factors provided four basic conditions in the experiment, which was then overlaid by the additional factor of experience. Approximately equal numbers of people from the nuclear industry and the public were presented with these four variants of the design. The basic design is summarised in Table 12.1.

In this design cell (a), where respondents come cold to an evaluation of options, and know the names of those options, might be seen as closest to

Table 12.1 Experimental design

ORDER EFFECTS	INFORMATION EFFECTS	
	LABELLED OPTIONS	UNLABELLED OPTIONS
Option Evaluations First	(a)	(b)
Option Evaluations Second	(c)	(d)

the usual or 'normal' mode of public assessment. There is not much possibility of people judging the attributes separately, and hence of combining them formally in the subsequent evaluations. Cell (d), however, is much closer to a formal evaluation procedure. It is a 'blind' assessment where respondents are constrained to judge options solely in terms of the technical estimates of impacts, and their assessment follows an explicit procedure for separate attribute evaluation. Cells (b) and (c) are intermediate in this progression.

When the index of predictability was entered as a dependent variable in this design, it was expected that Experience, Labelling and Order would all have effects, and that predictability would increase from a minimum in cell (a) to a maximum in cell (d).

Results

The sample of respondents was not chosen to be broadly representative of the general public. Indeed it was an aim of this part of the research to gather data from people who might be expected to have either considerable experience with radiation problems or, alternatively, to be at least fairly able with respect to the handling of complex data. Thus, for example, amongst the industrial group of 125, most of whom had considerable experience in the industry, including waste management, there were 29 radiation protection specialists. In the public group of 97 there was a disproportionate number of well-educated and highly qualified professional people.

The predictability of the individual evaluations was statistically adjusted to take account of the degrees of freedom lost because of the small number of evaluations (12) compared with the number of attributes (7). The mean values of this adjusted index are displayed in Figure 12.1.

Analysis of variance confirmed a main effect for Experience ($F1,211 = 8.0$, $p < .005$). For the public group there was no effect due to Order, but it was subject to an effect of Labelling ($F1,94 = 8.6$, p.005). The industrial group

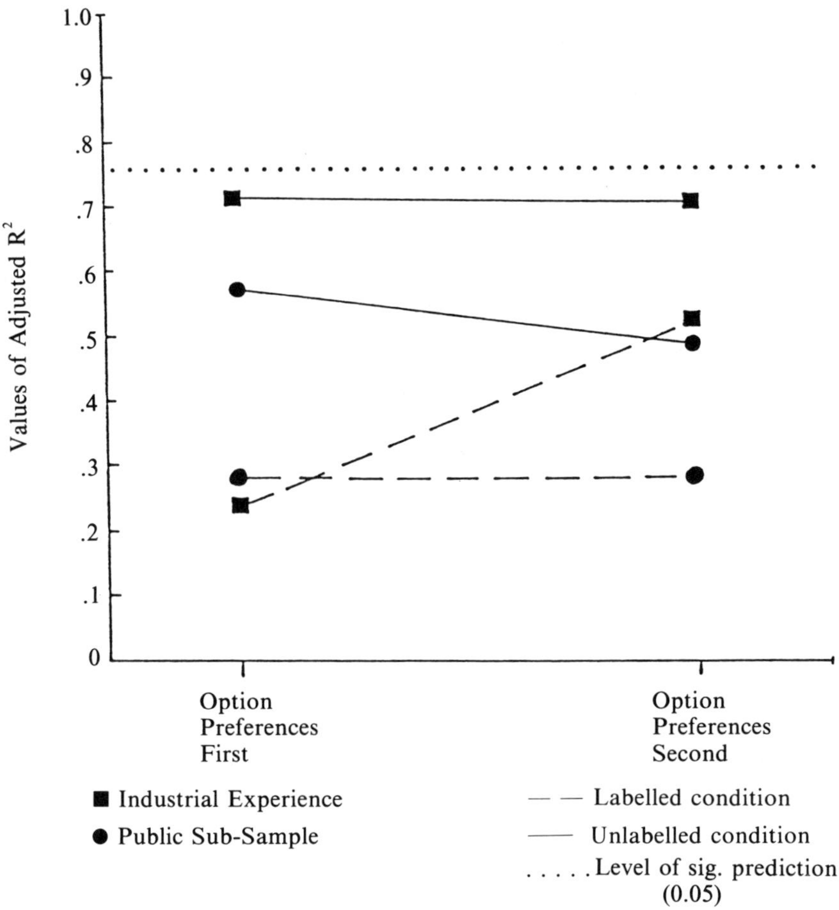

Figure 12.1 The predictability of option evaluations

was subject to main effects of Order (F1,120 = 3.9, $p<.05$), and Labelling (F1,120 = 23.2, $p<.001$) and to an Order x Labelling interaction (F1,120 = 4.9, $p<.05$). In total the experimental conditions accounted for nearly 9 per cent of the variation in predictability of the public responses and nearly 22 per cent of the variation of the responses from those in the industry.

A summary of these results is as follows:

1 When given practice with separate attribute evaluation, the public's predictability was not affected.

2 When those with industrial experience were given attribute evaluation practice, their predictability improved by a small amount (6 per cent).

3 For all those respondents limited to attribute information, practice with attribute evaluation had no effect.

4 When limited to attribute information, those with industrial experience gave more predictable option assessments than did the public.

5 The effect of separate attribute practice on those from the industry who were told the names of the options was to improve their predictability to the level achieved by those from the public who had been limited to attribute information only.

6 When respondents from either the public or the nuclear industry judged the options first, and had names and descriptions of those options, as might be the case in a simple public debate, the predictability of their responses was the same and was very low.

There is one caveat to these results. As the interaction in Figure 12.1 shows, the Order effect was limited to the Labelled condition (for the industrial group). Because the predictability in the Unlabelled conditions was already quite high, it is possible that improvements due to Order could not register in those conditions due to this 'ceiling effect'. This weakens the value of finding number three above.

However, added to the detail in Figure 12.1 is a horizontal dotted line. This represents the approximate value required in the index of predictability to reach statistical significance at the conventional level of $p < .05$. The line indicates that most respondents achieved levels of predictability in the use of attribute information rather less than that which would be expected by chance. Since the values in the Figure are mean cell values, the dotted line also indicates that less than half of the most predictable group, the industrial Unlabelled condition, achieved levels of predictability significantly above chance.

Some Conclusions

There are powerful messages in these results for those concerned with the response of the public to technological innovations. They show the strong effects associated with the difference between the 'normal' and 'normative' modes of evaluation. They also confirm that people with considerable experience in the industries concerned are more systematic when using technical information, as might be earnestly hoped, and that such people are open to training effects with a more formal procedure, such as separate attribute evaluation.

However, it is clear that the systematic use of technical information is no different between those experienced people and the public when the objects of appraisal are identified by name and it is possible for people to import their wider knowledge, values and beliefs to assist them in those appraisals.

To underline the meaning of these findings note that the huge fall in predictability associated with Labelling means that public and experienced technologists alike *chose not to use* the numerical impact information, at least in a way which can be modelled with the otherwise robust linear technique, when they 'knew' what they were evaluating.

Of course it could be argued that the technologists simply 'knew better', and that once they were aware of the nature of the options they were able to judge them more effectively – by ignoring the impact data and making use of their wide range of other relevant valid information. In the case of the public, who generally would lack such additional detailed knowledge, the reduced predictability would simply look like prejudice.

But this argument has a hollow ring. Even when constrained to use only the technical information on 'unknown' options, many of the technologists failed to use that information in a highly systematic way, and it must be more likely that the human penchant for categorical judgement outweighed the effort involved in careful evaluation of the impact data.

If this is the case for highly qualified and extensively experienced technologists, then the more so for the public. Clearly, the mere provision of additional purely technical data, and/or the 'education' of the public, would not address such effects.

References

Barraclough, L.M., and Mobbs, S.F., 1988, *Indicative Radiological Impacts and Costs of Radioactive Wastes*, NRPB Working Paper, Oxon

Brown, J.M., and White, H.M., 1987, 'The Public's Understanding of Radiation and Nuclear Waste', *Journal of the Society of Radiological Protection*, 7(2), pp 61–70

Brunswik, E., 1956, *Perception and the Representative Design of Experiments* (2nd edn), University of California Press, Berkeley

Edwards, W., 1965, 'Optimal strategies for seeking information: models for statistics, choice reaction times, and human information processing', *Journal of Mathematical Psychology*, 2, pp 312–29.

Edwards, W., 1977, *How to use Multi-Attribute Utility Measurement for Social Decision Making*, IEEE Transactions of Systems and Management Cybernetics, SMC-7, pp 326–39.

Hammond, K.R., and Adelmann, L., 1986, 'Science, Values and Human Judgement', in Hal R. Arkes and Kenneth R. Hammond (eds) *Judgement and Decision Making*, Cambridge University Press, Cambridge

Hammond, K.R., and Summers, D.A. 1972, 'Cognitive Control', *Psychological Review*, 79, pp 58–67.

Hart, D., 1988, *Weighting Factors for Cost and Radiological Impact for Use in Comparisons of Waste Management Methods*, NRPB, Oxon

Keeney, R.L., and Raiffa, H., 1976, *Decisions with Multiple Objectives: Preferences and Value Tradeoffs*, Wiley, New York

Lathrop, J.W., and Watson, S.R., 1982, 'Decision Analysis for the Evaluation of Risk in Nuclear Waste Management', *Journal of the Operational Research Society*, 33, pp 407–18

Lee, T.R. 1987, 'The public's view of the costs and benefits of alternative energy systems', *Proceedings of the Royal Society of Edinburgh, 92B*, pp 159–71

Lee, T.R., Brown, J., and Henderson, J., 1984, 'The Psychology of Nuclear Anxiety', in S. Surrey (ed.) *The Urban Transportation of Irradiated Fuel*, McMillan Press, London

Lindblom, C.E., 1964, 'The science of muddling through' in W.J. Gore and J.W. Dyson (eds) *The Making of Decisions*, Collier-Macmillan, London

Raiffa, H., 1968, *Decision Analysis*, Addison-Wesley, Reading, Mass.

Rawls, J., 1971, *A Theory of Justice*, Harvard University Press, Cambridge, Mass.

Simon, H.A., 1955, 'A Behavioural Model of Rational Choice', *Quarterly Journal of Economics*, 69, pp 99–118

Slovic, P., 1969, 'Analysing the expert judge: a descriptive study of a stockbroker's decision processes', *Journal of Applied Psychology*, 53, pp 255–63

Tversky, A., and Kahneman, D., 1974, *Judgement under Uncertainty: Heuristics and biases*, Cambridge University Press, Cambridge

van der Plight, J., Eiser, J.R., and Spears, R., 1986, 'Construction of a Nuclear Power Station in One's Locality: Attitudes and Salience', *Basic and Applied Social Psychology*, 7, pp 1–15

von Neumann, J., and Morgenstern, O., 1947, *Theory of Games and Economic Behaviour*, Princeton University Press, Princeton

von Winterfeldt, Detlof, and Edwards, Ward, 1986, *Decision Analysis and Behavioural Research*, Cambridge University Press, Cambridge

CHAPTER 13

Innovation and the Political Context of Technical Advice

Roger Williams

It seems important to begin with the general premises which underlie this chapter. There are four main ones.

1 Nation states in the developed world are, and will for the foreseeable future remain, in intense industrial and commercial competition with each other. This competition is shaped by government policies, led by private companies, and squarely predicated upon successful technology-based innovation. The competition is not new but its significance has greatly increased in recent decades, partly due to the avoidance of major war, partly due to the accelerating dynamic of technology. There are good reasons for believing that the United Kingdom is, even after the efforts of the last decade, not well placed in this competition.

2 Publics in the advanced industrial countries, though almost without exception attracted to the new and enhanced products which technical innovation makes possible, nevertheless evince concern about both the employment and the health and welfare implications of technical change, particularly as they see these implications bearing upon them. It is material in this connection that current generations are both better educated and have higher qualitative and quantitative life expectations than did previous generations: 'the overall consequences of all the technological progress of the past century has been to confine death very largely to the aged, whereas throughout all of human history death occurred more or less regularly among all ages of life' (Urquart and Heilmann, 1984).

3 The enormous differences in standard of living and in demands made on natural resources, especially energy and water, between the populations in advanced industrial societies and those in the less developed world are morally indefensible, possibly politically unstable, and certainly economically inefficient. Even so, only limited expectations can be entertained of the erosion of these differences in the immediate future.

4 Serious questions have recently been raised about the level of indus-

trialisation which the planet itself can sustain. These questions apply both generally and with respect to particular locations and instances of environmental damage. However, a substantial gap exists between public awareness of, and political rhetoric about, these several matters on the one hand and significant responsive action on the other. And here too it would be overly optimistic to expect any early and radical change.

Premises might be the wrong word for the above four points, but in whatever way they are described it is clear that technology is at the root of all four. Technology has no equal as a change-producing factor and it is therefore entirely appropriate that public acceptance of innovation, the ultimate embodiment of technology, should become a central concern of government (see, for example, Mills and Williams, 1986).

One can usefully observe here the distinction between 'acceptability' and 'acceptance'. The first is inherent in a particular technology or innovation itself, whereas the latter depends upon the reaction of those individuals and groups upon which the technology or innovation impinges. A second distinction might also be drawn: acceptability reduces ultimately to either/or, whereas acceptance implies a spectrum of states ranging from grudging toleration through to proselytising support. This point is not minor: the position reached on the acceptance scale can be expected to influence directly the penetration and success of those technologies which a given society judges to be acceptable.

And semantics having arisen, it is also briefly worth noting that 'public' too is a difficult word. Certainly, it is a function *inter alia* of distance (from the immediately affected to the distantly touched), the particular technology concerned, the media attention attracted, and so on. In addition, an uncertainty principle applies in assessing public opinion as in determining the position of an atomic particle: you cannot measure without affecting what it is you measure.

By contrast, the idea of innovation is more straightforward than either 'acceptance' or 'public'. The complexity in this case lies rather in the process. Innovation is typically grounded in a perceived correspondence between technical potential and manifest need or latent demand. It entails various stages from research (applied, strategic, or even basic) through development, to design, experiment, prototype and production, though these stages are not to be conceived as following each other linearly, nor need they all always be present. There may well be a place for invention and serendipity as there is more usually for organisation and system. Achievement and projection need to be continuously monitored against the environment in which the fruit of the innovation will eventually emerge. This environment may be static (as with the Apollo programme to put a man on the moon), or it may be changing independently (as with the air transport context in which Concorde finally emerged), or it may be changing in direct response to the innovation process itself (as with the American anti-ballistic missile project in the 1960s or the Strategic Defence Initiative in the 1980s). The innovation process usually proves to be highly dependent on a wide range of specialist skills, those of qualified scientists and engineers usually being seen as central but the contribution of systems analysts, stylists, production managers, marketing specialists, even patent lawyers and others also possibly being relevant. There is great scope for innovation to fail, or if it succeeds to disappoint in its scale of realisation. There is room for confusion, and therefore controversy, about the respective roles in innovation of free market forces as against supportive interventionism, or stimulating legislation, or intelligent purchasing by public bodies. There is also room for debate about the appropriate balance between co-operation on generics and competition on specifics. In short, innovation now has a large literature, and deserves it (Henwood and Thomas, 1984; *Oxford Review of Economic Policy*, 1988; and Ballard *et al.*, 1989).

Within this literature, it is readily seen that the public acceptance of innovation itself has many dimensions. The most basic and the most sensitive relates to the health and safety questions which it poses. These may be immediately explicit, or they may reveal themselves only over a long period and possibly after substantial research. It can be extremely difficult to delineate a threat in a way which is both statistically accurate and robust and also readily assimilated by those lacking a mathematical disposition. Another dimension concerns the more or less direct economic loss associated with new technologies, arising because jobs disappear, skills are devalued, capital is undermined. A third dimension, following Hirsch (1977), involves the disappearance of positional goods: the impact of the motor car perfectly illustrates this. The implication of new technologies for privacy constitute a fifth dimension and a sixth would be the modifications to social behaviour facilitated by new technologies and which attract criticism and controversy. Beyond this again are still deeper matters, as for instance the disappearance of a supposedly simpler life lived nearer to the grain of nature, or the sense that technology is somehow 'out of control' in a political world which that technology, it can hardly be denied, has made smaller, more interdependent, and less predictable.

There is now a substantial literature on the ethical and associated dilemmas involved in giving technical advice (Nelkin, 1975, and Golden, 1988). Almost thirty years ago, when the critical issue was nuclear weapons policy, Gilpin (1962, pp 15–19 and 334–40) was perceptive and direct: 'The advice of experts to the policy maker . . . is seldom if ever solely technical . . . careful analysis of its substance will reveal it to be political in nature . . . The necessity for such a combination of technical and political judgement gives to the scientist-adviser his special status . . .'. Gilpin believed that scientists needed to 'understand the nature of politics', and he even went so far as to outline a 'professional code for the scientist-adviser'. In the mid-1960s, dealing with broadly the same subject matter, Jacobson and Stein (1966, p. 479) could write that it called for 'great restraint and discipline to limit a judgement to the scientific element'. As they saw it therefore, a scientist could 'give a wide range of answers without violating his scientific integrity', these answers reflecting not only his own 'preferences and predelictions' but also 'the way in which the question is phrased and asked and the form and type of answer permitted'.

Others who addressed these issues in the mid-1960s were Bundy (1963) Schilling (1964) and Skolnikoff (1967). For Bundy (1963) it was of 'the highest importance' that a scientist should 'respect his calling', being careful to 'limit the occasions upon which he speaks ex cathedra, lest he lose his reputation for infallibility'. Schilling offered an inventory of 'six policy perspectives . . . moderately characteristic of many scientists . . . who have participated in [American] national security policy.' As to these opinions – 'naive utopianism or naive belligerency . . . the "whole problem approach" . . . quantum jumps versus improvements . . . technology for its own sweet sake . . . a sense for paradise lost . . . science serves mankind' – Schilling thought them neither bad nor good, but he was clear that non-scientists would be 'wise to be alert to them'. It was in reviewing the work of the President's Science Advisory Committee and the Special Assistant to the President for Science and Technology that Skolnikoff came to the conclusion that 'it is impossible to provide technical inputs on important issues in a political vacuum . . . to provide technical advice independently of the implications of that advice and the political variables would have been meaningless'.

At the end of the 1960s, the US Congressional Hearings on the anti-ballistic missile (ABM) programme led to a report by the Operations Research Society of America (ORSA, 1971), concerned that these hearings revealed a wide public disagreement between scientists. The Society was alarmed lest such conduct bring the practice of science, and operations research particularly, into public disrepute. However, many who had given evidence in the ABM hearings subsequently objected to the ORSA study, finding it 'absurd', in that 'the operations research aspects of the debate became inextricably linked with, and indeed probably less important than, judgements on . . . technical issues'.

In the early 1970s the journal *Minerva* published a major series of articles concerned with science advice. It was in the course of these that Alvin Weinberg (1972) advanced the concept of 'trans-science'. Weinberg argued that 'when the questions raised cannot be answered from existing scientific knowledge or from research which could be carried out reasonably rapidly and without disproportionate expense, then the answers must be trans-scientific and the adversary procedure seems therefore to be the best alternative'. Another contributor, Harvey Brooks (1972), agreed with this, but chose to distinguish between 'inherently unaswerable' and 'currently unanswerable' questions. Edward Shils (1972) was much more sceptical, since for him 'the situation of a scientific adviser is different from that of a lawyer arguing a case in court . . . There is a check . . . the judge . . . as much an expert as the lawyer . . . The situation of a scientist giving advice to a politician . . . is much closer to that of a physician in relation to the patient'.

Two other distinguished analysts in this field have been Dahl (1950) and Dror (1971a, 1971b). As early as 1950 Dahl put the onus on the technical adviser to discriminate wisely between what he called 'expert-authoritarianism', 'expert-neutrality' and 'policy-advice'. Dror has tried to delineate the 'very elementary mistakes' which can be made by scientists 'entering the domain of policy'. Dror finds the 'serious defects' in many pronouncements by scientists on policy questions 'inexcusable'. His justification for this is that he thinks it perfectly possible for scientists to present recommendations in formats which 'clearly distinguish between the different bases of the recommendations'. Dror identifies no less than eight shortcomings to which he regards science advisers as prone. Such people he thinks oscillate 'between naivety and cynicism', and between the extremes of confining themselves to narrow professional frameworks or freely going far beyond these. What he fears is a 'mix up between reliable factual knowledge, implicit axiomatic assumptions, provisional theories, conceptual taxonomies, doubtful hypotheses, and various types of hidden value judgements, such as on substantial goals, or willingness to take risks, and on evaluation of time'.

There is likely to be continuing disagreement as to the particular advantages and disadvantages of scientific advisers when dealing with trans-scientific questions. Pierls (1969) and Nichols (1972) demonstrate perfectly how widely divergent views can be on this subject. Pierls regards scientists as 'well qualified to make contributions to debates in which they are not in exclusive possession of relevant facts . . . scientists are capable of somewhat greater objectivity and detachment in discussing such problems . . . For public leaders . . . it is a perfectly normal course of action . . . to hide or misstate ascertainable facts'. Nichols on the other hand believes of scientific advisers that 'while their advice on means is as authoritative as their knowledge permits, their views on ends are usually no more valid, and possibly less representative of public preferences, than those of elected or appointed officials and non-scientists'.

It can be argued that trans-science is a special case of a general problem, because another way of approaching trans-science is to recognise that when one cannot make deductions and predictions strictly according to scientific laws, then the only statements which can be made about the future are probabilistic ones. Policy advice is effectively predicated on two assumptions and a preference. The first assumption is that the future contains genuinely different possibilities; the second that a clear connection exists between different courses of action now and outcomes at some time in the future. The preference ranks those outcomes, and therefore the courses of action available now.

There are still considerable difficulties. How, for example, is one to assign a numerical probability of occurring to an event which has never yet in practice actually occurred and about whose mechanism of occurrence there is substantial room for argument? Or how are different groups, possibly even different generations, to be included in a preference calculus? And how confident can one be that all possible courses of action have been determined, all independent but relevant events recognised, all interconnections traced, all outcomes faithfully rendered? Nor does this exhaust the possibilities.

The wish to see ahead more systematically has more than once received institutional embodiment in recent decades. A classical example is the Office of Technology Assessment founded by the US Congress in 1972. This has now spawned many imitators. The OTA is in the business of relating developments in science and technology to options in society, economics and politics. The OTA and bodies like it could hardly be said to have revolutionised the methodology for analysing the future but they have made such exercises intellectually respectable. Their reports may be heavier going than Verne, Wells, or Nostradamus, but they are more likely to influence the climate of thinking. They may even therefore have something partially self-fulfilling about them. In any case, almost by definition, they are largely confined to 'surprise free' extrapolation.

The hard umbra of unequivocal fact in any complex situation is commonly relatively small in relation to the penumbra of more or less well-informed, and more or less well-meant, opinion. The corollary is that whereas differences over matters of fact are in principle readily decided, differences over matters of judgement are rarely open to any simple test procedure. More specifically, given that decisions in the real world have normally to be taken in the face of much uncertainty, this uncertainty may be technical or economic, or it may relate to social or political behaviour. Commonly it is a mixture of all these forms. Four examples provide useful illustration.

In the case of thermo-nuclear fusion the first uncertainty is purely technical: if the technology were proved there might or might not be further uncertainties, for example in regard to the comparative economics of fusion and fission power, but serious discussion of such further uncertainties must really wait upon the resolution of the technical one. In the case of tidal power, say specifically the Severn Barrage, the technical uncertainties are negligible but here the economic ones loom large, which again is not necessarily to say that, were the economic issues set aside, the socio-political questions could then be taken for granted. And finally, it is interesting to compare the Third London Airport with the Channel Tunnel. Both of these were long-discussed projects but in the end in the latter case all objections, technical, economic and political, were overcome, at least sufficiently to allow boring to get underway, whereas in the other, political objections eventually thwarted development plans.

Any policy issue has, at a given time, an informational state. This describes what is known about that issue, what is not known, what search or research is underway to shift the balance between the known and the unknown, whether the information is widely available or is confined to particular individuals or organisations, and so on. The informational state can then be used, by inference, projection or otherwise, to identify policy options, and more recently to generate a technology assessment, an environmental impact statement, or similar report. The main aim of such reports is to reduce uncertainty. It may be possible to deploy various economic techniques, capital appraisal methods or cost/benefit or cost/effectiveness analysis for example. Risk analysis may be relevant. The results or scenarios may be illuminated by sensitivity studies, when underlying assumptions are modified and their effects traced. Opportunity costs may be deduced, or wider implications be identified.

These various procedures both take time and occasion expense. In principle, the investment made in them would reflect in some way the scale of the consequences foreseen and the associated uncertainties, but this is a counsel of perfection, since inevitably consequences are never completely perceived, there are commonly independent time and money constraints, and institutional restraints and biases will typically make their presence felt.

Although there will normally be many more specific complications in any individual case, this in outline is the context in which advisers advise. It is, in short, the country of the blind but despite his one eye the adviser nevertheless cannot expect to be king. The adviser is in fact in a highly political situation.

Specialist advice is conveniently regarded as one of three main kinds of input to decisions, non-specialist administrative and political advice being a second (and more traditional) type, and the communication of pressures, of varying degrees of legitimacy, the third (and no less traditional) type. 'Pressures' here subsumes the effects of ideology, mandates, promises, debts and so on. A distinction is also being made here between 'pressures', taken to be self-interested or 'partial' information, and 'advice', both specialist and non-specialist, classically assumed to be more neutral or 'balanced' information, though in the real world advice is rarely of this unblemished sort. It is also essential to remember that decision makers have their own biases, whether latent or manifest. They are also normally dealing with many pressing issues simultaneously, whereas an adviser may be dealing with only one. Furthermore, counsel, as Chaucer said, is no command. In the end, pressures, specialist and non-specialist advice, and bias, can all be reduced to a mixture of three elements: what the existing situation is understood to be, and what it is thought possible, and what it is thought desirable, to do from that situation. And in the end too, the most telling question one can ask about any advice, specialist or otherwise, concerns the use to which it is put, or not put, though only occasionally will it be possible unequivocally to answer this question.

Ideally, technical advice would be clear, objective and relevant. Its provision would be timely and it would be accorded an importance in decision making which as precisely as possible reflected its significance. When, as so often happens, practice falls short of this ideal, the responsibility may be either that of those who advise, or the problem may lie with those who receive advice, or the deficiency may derive from the character of the advice itself, or the principal difficulty may be the fault of the structure in which the advice is formulated.

A very common shortcoming with advice is that it may be too vague, not specific enough to point to a definite course of action. No-one has expressed better than Fisher the advantage held by those who deal in specifics. 'Within a bureaucracy', he asserts, 'those who have worked out a specific plan and come forward with a yesable proposition are likely to carry the day' (Fisher, 1971, p 27). In the scientific context the point was put somewhat more cynically by Snow in his account of the Tizard-Lindemann affair. Snow stressed: 'the prime importance, in any crisis of action, of being positive what you want to do and of being able to explain it. It is not so relevant whether you are right or wrong . . . But it is cardinal that you should be positive' (Snow, 1963). Political decision making in other words, is about real choices in real time, and not about general philosophy, and an adviser must be prepared to be faced with a demand for a clear opinion.

A closely related point was made by Eiduson (1966). For her the most important consideration is the amount of time spent in proximity to decision makers, 'being present at the critical time', and learning through experience. The advisers she questioned also spoke highly of abilities such as those 'to make decisions without worrying about too many details . . . to live with the consequences . . . to arrive at solutions to problems now'. But if these are desirable attributes they also have their costs. For example, the more time one spends close to decision makers, the less time one spends as a scientist, and those who concentrate on arriving at solutions to problems 'now' risk losing their strategic grasp. The opposite problem also frequently arises: access to the actual point of decision may be limited in any one of several ways.

As well as being 'yesable', advice must also make political sense. Zuckerman has underlined the constraints on decision makers which stem directly from the situation they inherit. The value of counsel for him varies 'according to the circumstances of the moment and the executive responsibility of the authority which is being advised'. More bluntly, in Huitt's (1968) words, 'what is most feasible is what is incremental'.

Most decisions manage only to approximate rationality, and advice can easily become part of a self-reinforcing orthodoxy. Committees are notorious for this, but advisers as individuals are also often guilty. The adviser will be influenced to at least some extent by his perceptions of his own circumstances and his expectations, while those he advises will be mindful of his previous track record and possibly also his soundness. And there need be no correlation between the professional qualities which led to an individual becoming an adviser and those he requires in the policy arena.

Manipulation of advice is another problem. The balancing of pressures and non-specialist advice is a standard political technique and lay decision makers may not always appreciate that treating specialist advice in the same way can lead to error or worse. In the worst situation, the adviser may simply be required to support, on technical grounds, a position preferred by those he or she advises on non-technical ones. Sometimes, on the other hand, it is the adviser who may be doing the manipulating, as in the sort of circumstance described by Jones (1972): 'Timing can sometimes have a crucial effect. A new point introduced at a late stage can produce an effect quite disproportionate to its merit . . . I have seen committees, even up to Cabinet level, almost stampeded by isolated pieces of "stop press" information.' Mandelbaum (1969) has made the same point more generally: 'The more serious obstacles to mature consideration are the crisis atmosphere (often artificial), the fragmented machinery . . ., and the momentum of technological impulse.' Very often events will exert their own pressures, the option of action becoming a need for reaction, or crisis otherwise precipitating its own urgency. Somewhere along the way matters may become personalised, or stock constituencies demand to be placated. There will invariably be frustrations as innocence, or at least protested rationality, confronts the inefficiencies, inequities, delay, noise and confusion of the policy process.

Sometimes it is possible to redefine a problem, or to postpone a final decision. Redefining a problem, when this is a possible strategy, can be a transforming experience, in extreme cases removing the need for a decision, exposing paths not previously perceived, or reordering the weighting attached to options and outcomes. The desire to wait before making a decision until the evidence is clearer, or until a consensus between key parties emerges, is very understandable. This may even be the best, or indeed in some instances only, policy. In other cases, however, it may add to the costs, financial and otherwise, which have in the end to be borne. Understanding the situation one is in, and knowing the right time to come to a decision is, in the techno-political area as in others, central to the art of decision making.

Yet another complication may follow from the mismatch between the quantitative and qualitative dimensions of advice. Sir Richard Clarke is among those who have pointed out that 'measurable factors are not necessarily as important as those which are not'. But like others he has also noted the risk that a decision 'will be dominated by prejudices and that the less important intangibles will be given much more weight than they are worth'.

To sum up, expert advice is a political weapon, 'a well-recognised ploy', as Don Price (1965) described it a quarter of a century ago. And it is hard to improve on Bryson's advice, offered even longer ago: 'a kindly scepticism regarding the motives of all men in practical affairs is the usual attitude of experienced men of action: it becomes the expert quite as well if he can attain it'. Bryson noted that in political disputes as against scholarly ones it was 'a man and not an opinion that wins', and he reminded the adviser that 'the one who makes the final policy decisions . . . has a primary task in keeping his own power as against rivals . . . exercising power is drastically interfered with by the need to hold it' (Bryson, 1951).

Conclusions

Seven broad conclusions are put forward here, not in any belief that they are the only appropriate ones to draw, but because they are all self-evidently important and because consideration of them must at least prompt insight and constructive reflection.

1 Society at large needs to be confident that the effects of technology, including its injustices, are not being permitted to happen casually. The public needs to be reassured that both the economic and the safety consequences of technology are being monitored and as necessary addressed by a concerned, competent and just state. Where regulation is required, this must be seen to be clearly separated from sponsorship, the regulators must have the necessary powers, and the public must be able to satisfy itself that regulation is succeeding. The fact that the economic consequences of technology are not incident evenly on populations creates an onus to ensure that there are adequate compensation and support schemes for those unavoidably disadvantaged by new technical development which, seen in the round, confers worthwhile social benefit. And the possibility should always be considered of financing compensation and support schemes out of the profits generated by the technical change which causes the displacement. Retraining has its place here and initial education should henceforth provide increasingly for later flexibility. The first industrial revolution was accompanied by ugliness and despair and these features remain all too common. They are a brake as well as a reproach which successfully innovating societies do not need and cannot afford. Broad social fairness is not incompatible with high rates of technical change.

2 It is desirable to promote in as high a proportion of society as possible a facility with technology – not simply an understanding or toleration but real facility. The engineer does not see technology as remote and threatening but as something to be employed for betterment and in respect of which there are skills to be learned. Good technology education is about attitudes to innovation but also about the practical aptitudes needed to realise such innovation. Technology education needs to be seen as supplementing and not substituting the values of a traditional education, though it is purposeful in a way much science education is not. There

is no suggestion here of indoctrination, rather the meeting of a fundamental requirement in advanced industrial society, something which many, it may even be most, in the UK have still, painfully, to learn.

3 The reward and status systems in society need to be evaluated from the perspective of their influence on the processes of innovation. Are individuals in their private and career contexts encouraged to be entrepreneurial or conservative? Are the incentives to improve performance undermined by a cultural conformity which tends to favour achieving security? Do the mores of the financial and managerial worlds encourage a short- or a long-term view? All these questions are directed at the degree of congruence which exists between the requirements of technologically dynamic and successful innovation and the historically rooted values of contemporary societies.

4 Society and the organisations within them both respond better to clear goals. In democratic societies the support of electorates is needed to underwrite the view that the steady and responsible advance of technology is the key to national, communal and individual betterment. New technical ideas are best pursued with high variety and imagination, but also with discipline, including financial discipline. The aims are a national pride which stops short of chauvinism together with high organisational *esprit de corps*. There is now much evidence to show that some societies and some organisations are much better at these things than are others.

5 There is scope for substantially more quantitative underpinning of many more public policy issues than presently occurs. For example, public and private R and D statistics could be much more comprehensive, less subjective and more internationally comparable than is currently the case; better data on the failure characteristics of mechanical components and complete artefacts could enhance reliability, safety and quality; public health policy would benefit greatly from improved epidemiological information; and so on. While it is true that acquiring information imposes its own costs it is not clear that societies are as yet near the ceiling of reasonableness in this respect.

6 There is no difficulty in arguing for advisory and decision-making structures as open as possible to scrutiny by the public and its agents. The correlation between openness and accountability is direct and accountable governments are much to be preferred. It is a fact of life that most decision makers and advisers would rather that full accountability applied to everyone but themselves. Granted, there is also room for argument that democratic government is not always best served by openness at all times and in all circumstances. However, it is not irrelevant here that science itself has advanced precisely because

of the openness of its processes and closed science has a demonstrated tendency to be driven down in quality. Nor is this conclusion undermined by practices adopted because of national or commercial security. It is, however, not widely accepted that what holds true for the progress of science obtains no less in the case of professional scientific and technical advice. On the contrary, the prince has commonly relied upon only one expert adviser, and not infrequently with the deliberate aim of preventing the emergence of differences between experts. On matters which rely largely on expert opinion, the wise non-expert decision maker may however see advantage in encouraging more or less open debate between specialists, in that such debate should at least ensure that all approaches are considered and the decision maker protected from later criticism. The reasonableness of this approach is itself not unattended with difficulty, in that the more technical debate they encourage the more non-expert decision makers may eventually come to be seen as themselves deciding between technical arguments. Nevertheless, they above all should reflect that *post factum, nullum consilium.*

7 The technical character of many contemporary public questions leads to the thought that non-expert decision makers might with advantage aim to acquire an enhanced technical mindedness, a 'feel' for technical questions similar to that they would claim for non-technical ones. Even an elementary technical mindedness would avoid many of the absurdities of which non-expert decision makers have been guilty in recent years. But here too there is a possible cost, that which comes from the possession of a little knowledge without a corresponding humility.

What presently seems quite impossible, whether or not desirable, is any initiative to slow down the contemporary rate of technology-led innovation. However much one might yearn for a more slowly evolving environment, for the foreseeable future Francis Bacon's words can be expected to have only a romantic relevance:

> It were good, therefore, that men in their innovations would follow the example of time itself, which indeed innovateth greatly, but quietly and by degrees scarce to be perceived.

References

Ballard, Steven *et al.*, 1989, *Innovation Through Technical and Scientific Information: Government and Industry Cooperation*, Quorum Books, New York

Brooks, Harvey, 1972, *Minerva*, X, 3, July, p. 484 (letter)

Bryson, Lymen, 1951, 'Notes on a theory of advice', *Political Science Quarterly*

LXVI, September, 3, pp 321–39, at pp 329–32
Bundy, McGeorge, 1963, 'The Scientist and National Policy', *Science*, 139, 1 March, pp. 805–9
Dahl, Robert A., 1950, *Congress and Foreign Policy*, W.W. Norton Co., New York
Dror, Yehezkel, 1971a, *Ventures in Policy Sciences*, American Elsevier Co. Inc., New York
Dror, Yehezkel, 1971b, *Design for Policy Sciences*, American Elsevier Co. Inc., New York
Eiduson, Bernice T., 1966, 'Scientists as Advisers and Consultants in Washington', *Bulletin of the Atomic Scientists*, October, pp.26–31
Fisher, Roger, 1971, *Basic Negotiating Strategy*, Allen Lane, The Penguin Press, London
Gilpin, Robert, 1962, *American Scientists and Nucear Weapons Policy*, Princeton University Press, Princeton, NJ
Golden, William T., 1988, *Science and Technology Advice to the President, Congress and Judiciary*, Pergamon Press, New York
Henwood, Felicity, and Thomas, Graham, 1984, *Science, Technology and Innovation: A Research Bibliography*, Wheatsheaf, Brighton
Hirsch, Fred, 1977, *Social Limits in Growth*, Routledge and Kegan Paul, London
Huitt, Ralph K., 1968, 'Political Feasibility', in Ranney, Austin (ed.), *Political Science and Public Policy*, pp.263–75, Markham Publishing Co., Chicago
Jacobson, Harold Karan, and Stein, Eric, 1966, *Diplomats, Scientists and Politicians*, The University of Michigan Press, Ann Arbor
Jones, R.V., 1972, 'Temptation and Risks of the Scientific Adviser', *Minerva*, X, 3, pp.441–51
Mandelbaum, Leonard, 1969, 'Apollo: How the United States Decided to go to the Moon', *Science*, 163, 14 February, pp. 280–94
Mills, Stephen and Williams, Roger (eds), 1986, *Public Acceptance of New Technologies*, Croom Helm, London
Nelkin, Dorothy, 1975, 'The Political Impact of Technical Expertise', *Social Studies of Science*, Vol. 5, No. 1, pp. 35–54.
Nichols, Rodney W. 1972, 'Some Practical Problems of Scientist-Advisers', *Minerva*, X, 4, October, pp.603–13
ORSA, 1971, 'Guidelines for the Practice of Operations Research', *Operations Research*, XIX, 5, September, pp. 1123–58.
Oxford Review of Economic Policy, 1988, Vol. 4, No. 4, Winter special edition, *Technical Progress*
Pierls, Sir Rudolf E., 1969, 'The Scientist in Public Affairs: Between the Ivory Tower and the Arena', *Bulletin of the Atomic Scientists*, November, pp. 28–30.
Price, Don, 1965, *The Scientific Estate*, Harvard University Press, Cambridge, Mass.
Schilling, Warner R., 1964, 'Scientists, Foreign Policy and Politics', in Gilpin, Robert, and Wright, Christopher, *Scientists and National Policy Making*, Columbia University Press, New York, pp.144–73
Shils, Edward, 1972, 'The Obligations of Scientists as Counsellors', *Minerva*, X, 1, January, pp. 107–10.
Skolnikoff, Eugene B., 1967, *Science, Technology and American Foreign Policy*, The MIT Press, Cambridge, Mass.
Snow, C.P., 1963, *Science and Government*, New English Library, London
Urquart, J. and Heilmann, K., 1984, *Risk Watch*, Facts on File Publications, Bicester, New York.

Weinberg, Alvin M., 1972, 'Science and Trans-science', *Minerva*, X, 2, April, pp. 209–22

Zuckerman, Sir Solly (later Lord), 1970, *Beyond the Ivory Tower, The Frontiers of Public and Private Science*, Weidenfeld and Nicholson, London.

CHAPTER 14

Entrepreneurship and Innovation from an Environmental Risk Perspective

Gerd Rainer Wagner

The Problem

Every technical innovation that entrepreneurs introduce entails social and political, economic and environmental effects. If these effects are uncertain in occurrence or extent – and this is typically the case – every entrepreneurial decision on an innovation creates certain social and political, economic as well as environmental risks. This chapter focuses specifically on the relation between technological entrepreneurial innovation and environmental risks. In this context the 'environment' means the 'ecological environment'.

The relation between entrepreneurial innovation and the environment can be studied under various aspects. On the one hand, the focus of study may be the environmental impact assessment of a proposed innovation. This is the problem already referred to. In this case the problem of public acceptance is uppermost particularly if this assessment is integrated into a formal licensing procedure organised by the state. On the other hand, one may ask how environment-related social or political requirements influence innovation by companies, and more precisely their perceived innovation risk. In this case public acceptance is only an indirect question, though no less relevant. Then the question is: how does advance acceptance of innovations, rather than simply reactive acceptance, influence the entrepreneurial risk and the environmental risk in general?

In the following I will concentrate on this latter question. I will study the relation between the entrepreneurial innovation risk and the environmental risk paying particular attention to the question of whether and how entrepreneurial decisions can influence or shape this relation over the course of time. I will refer to some concrete and current examples, starting with a simple model.

The Model of the Public Acceptance Line

There are, of course, many different forms of social advance acceptance of environment-related innovation risks. These are, for example, expressed in

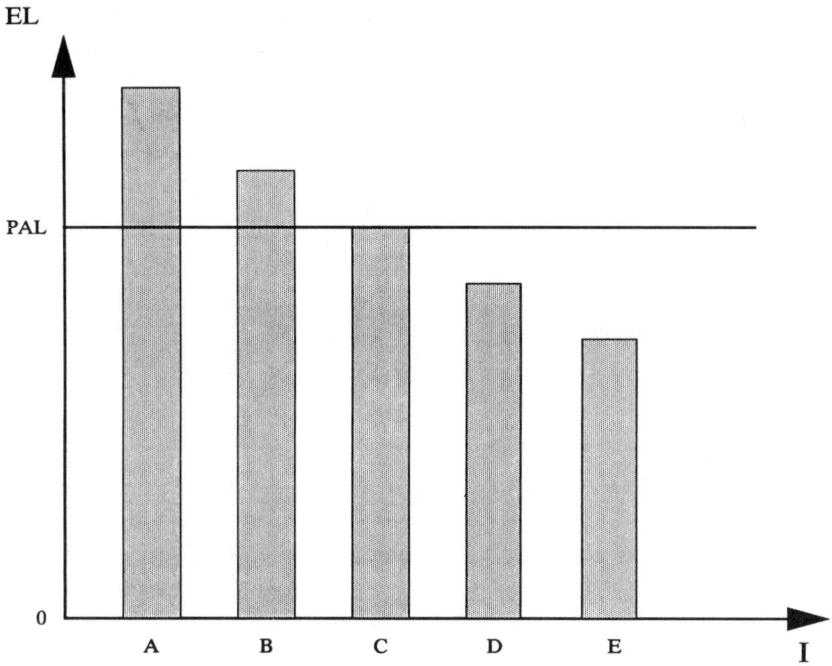

$0 \leq EL \leq PAL \longrightarrow EeR = 0$

$EL > PAL \longrightarrow EeR > 0$

I : Innovation Alternatives (A,B,C,...)

EL : Emission Levels of Innovation Alternatives

PAL : Public Acceptance Line

EeR : Entrepreneurial Environmental Risk

Figure 14.1 The model of public acceptance line

quite specific standards which must be observed by companies, such as standards for radioactivity in the surroundings of nuclear power plants, standards for the use of pesticides in agriculture, or standards for pollutant emissions by industrial production processes.

Standards of this kind always define the borderline between general, that is social, environmental risks and individual company risk. This is demonstrated in Figure 14.1 for the event of a public standard being set for high-emission production processes.

It is assumed that the company is faced with the alternative process innovations A to E in its production field. Measured on the ordinate, these alternatives represent different levels of air, soil, or water pollution. If the public authorities now set – in whatever way – a certain standard for possible emissions this standard can be termed the 'public acceptance line' (PAL). For all innovation alternatives whose emission levels do not exceed PAL (i.e. C, D and E) the 'entrepreneurial environmental risk' (EeR) is thus automatically zero. For all others (A and B) it is, however, greater than zero. In other words, the definition of a 'public acceptance line' delimits the field of possible transformation of external environmental risks into internal, that is, entrepreneurial environmental risks.

This fact deserves detailed attention. For now it is easily possible to explain certain actual phenomena.

Consequences of Changes in the Public Acceptance Line

Figure 14.1 describes the situation in which a certain innovation alternative is to be selected at a certain time, namely at t_0. It will be assumed that the entrepreneur selects E, because on the one hand, he or she wishes to achieve an entrepreneurial environmental risk of zero, and, on the other hand – possibly due to special environmental responsibility – it is the lowest possible emission level below the public acceptance line. Without the 'time' aspect the selection problem would thus be solved.

However, the specification of public acceptance lines characteristically changes over time, either due to more stringent public standards or to certain policy-induced market reactions. The tightening of public environmental standards is common practice, for example as a result of improved pollution measuring methods or intensified public pressure. From an entrepreneurial viewpoint such changes generally remain unproblematic as long as they take place in foreseeable steps and with due regard to the estimated pay-off periods of innovative investments.

Public acceptance line modifications resulting from certain policy-induced market reactions can be problematic, however. A paradigm case is the 'existing state of technology' standard as prescribed by the German Federal Immission Control Act. This standard gives rise to public pressure on investment since it can force companies to adopt innovations from others in the construction and operation of plants subject to licensing, provided that such innovations lead to lower specific emission levels. The idea of this standard is to dynamise the public acceptance line as a function of the general environment-related progress of innovation, namely by entering into entrepreneurial innovation competition.

In reality, however, specifying the 'existing state of technology' as a standard leads, in many cases, to significant collective rejections of entrepreneurial innovation and investment. For the individual entrepreneur such a

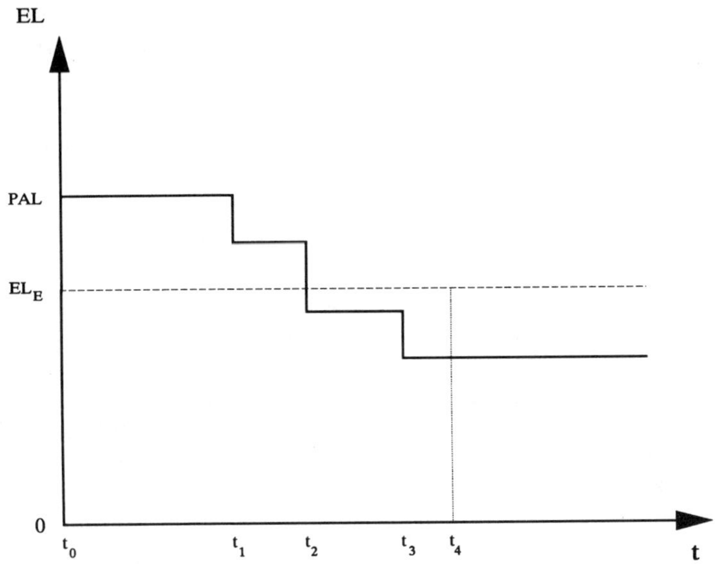

$0 \leq t \leq t_2 \rightarrow EeR_E = 0$

$t > t_2 \rightarrow EeR_E > 0$

EL_E : Emission Level of Innovation E

EeR_E : Entrepreneurial Environmental Risk in Case of Innovation E

$\overline{t_0 t_4}$: Pay-off Period of Investment E

Figure 14.2 Public acceptance line and pay-off period (case 1)

standard always involves the risk of public non-acceptance of an investment just completed as soon as an environmentally more effective innovation appears in the market. In other words, such a standard can endanger the individual entrepreneur's anticipated profits due to a sudden end of his temporary innovation-induced market monopoly. It can even cause losses if this standard prevents the necessary pay-off periods from being reached.

Figure 14.2 demonstrates this case, for if an entrepreneur has selected alternative E of Figure 14.1, the emission level of this alternative is expressed by the horizontal line EL_E in Figure 14.2. The pay-off period estimated for the investment is t_0t_4, the pay-off point is t_4. With the level of the public acceptance line given in t_0 constant over time, the environment-related entrepreneurial risk would be zero for the whole period, and thus would not inhibit the innovation.

However, with expected competing innovations and thus a consequential lowering of the public acceptance line in t_1, t_2 and t_3 a special entrepreneurial risk would start at t_2 and this risk would further increase in t_3. In concrete terms, at t_2 a loss risk would arise due to the public investment pressure since the pay-off point has not yet been reached. And as this loss risk is evident to all potential innovators when the changes of public acceptance line are formulated in this way, they will jointly develop defensive strategies to stabilise the initial level of the public acceptance line over time.

In these circumstances it proves to be collectively economically rational to refrain from actual innovation activities. And instead of achieving an improvement of the environmental situation through increased innovation competition, a reduction in competition and, at best, a stabilisation of the status quo of the general environmental risk will occur.

This effect is well known, and political counter-strategies are currently being developed, for example in the form of targeted public funding of specialised innovations to crack such cartels of refusal. But I shall not to go into further details on this particular subject.

Instead, I would like to consider another form of public advance acceptance. This alternative strategy does not specify a certain generally applicable environmental standard. Instead, this strategy aims at concrete innovative technological solutions and prescribes their general application or – what may lead to the same result – promotes such technologies exclusively with massive subsidies. Funding of the catalytic converter technology for automobiles in Austria and Germany is an appropriate example.

Publicly Specified Technology and Innovation Competition

If we return to Figure 14.1 and assume that all the innovation alternatives A to E are suitable to solve a certain technological problem, this alternative policy strategy consists in choosing one of these alternatives which is then declared obligatory or at least worth special public promotion by the state. It will be assumed that the public authorities select – consistent with the general standard of public acceptance line emission level, which would otherwise be set – alternative C. This will result in the following consequences.

With the obligatory specification of a certain technological solution the state itself takes on specific duties. By contrast with the specification of emission standards, the state will have to consider more fully the necessary pay-off periods of the investors to ensure that the companies will accept its technological specifications at all. For the individual innovator this leads to a reduction (or, even better, a removal) of his pay-off risk. This is shown in Figure 14.3 in comparison to Figure 14.2 for innovation E. (In comparison to Figure 14.1 it must be borne in mind that, though the public acceptance line in Figure 14.3 is formally identical, the respective specification is absolutely different.)

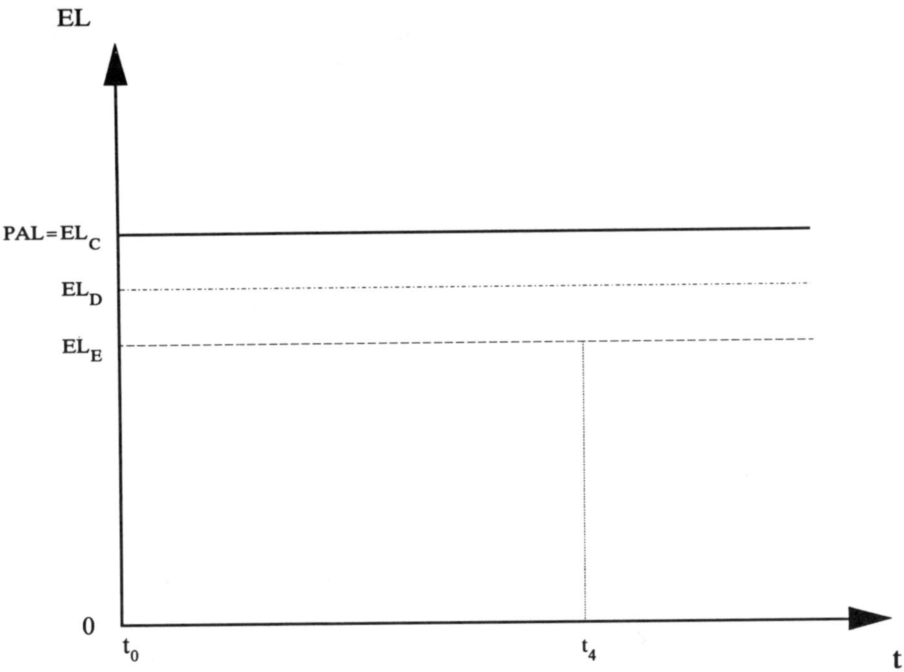

$$t \geq 0 \longrightarrow EeR_E = 0$$

$EL_{C,D,E}$: Emission Level of Innovation C, D, E

Figure 14.3 Public acceptance line and pay-off period (case 2)

As is also depicted in Figure 14.3, such a procedure also reduces (or even removes) the environmental entrepreneurial risk of this innovator E. From E's perspective this governmental strategy might therefore almost constitute the ideal case.

But such a solution is of economic benefit for innovator E only under quite specific and well-defined assumptions. Normally, a technological low-emission solution can only be attained at higher expense, in this case compared with C and D. Therefore, under these conditions the entrepreneur will only strive for solution E if the market is prepared to honour an innovation which promises lower environmental pollution with higher product prices. If this is not the case, innovation E would yield a smaller profit margin than C or D. Entrepreneur E would thus be facing a competitive disadvantage caused by a public decision, whereas entrepreneurs C and D would achieve a competitive advantage. If so, it would be economically rational for E (and also for D) to select investment alternative C too.

In other words, the specification of public acceptance lines based upon governmental decisions on technology can lead to a simultaneous reduction of economic and environmental risk for innovations, because of the generally accompanying commitments of the state. For this reason such an approach appears quite fascinating to many politicians and also entrepreneurs.

But the 'price' of such a way of action is normally an increased danger that the relevant innovation competition may be reduced significantly. As a result, further innovations may be generally excluded for an extended period of time – specifically as long as the public acceptance line remains unchanged – and thus further impulses for technological and environmental progress are suppressed. Again, the requirement of exhaust catalytic converters for cars could be quoted as an example.

Consequences for Environmental Cost and Risk Distribution

The public prescription of certain technological solutions can lead to further problematic consequences if one focuses on certain other fields. Take for example a case in which a company is forced to remove soil contamination attributable to its factory.

Consider the possibility that there are different technological processes available to solve the problem. They are identified by the respective 'levels of decontamination' (LoD) in Figure 14.4. In addition, the company anticipates the development of even more effective processes for the future, also expressed by LoD. Moreover, consider that the environmentally most effective method is usually the most expensive one.

Assume that, because of public opinion, the government compels the companies always to apply the environmentally most effective solution through legislation or other administrative instruments. In addition, assume – and that is in fact the actual case – in t_0 the public has no concrete knowledge of the latest innovations available or of the developments actually anticipated for the future.

Here the public will first attempt to let the state prescribe the application of a specific technology, which though known to the public might not be the most advanced. With increasing knowledge of actual technological possibilities and perspectives the public will then increase its requirements. From the viewpoint of the company in question this will cause a jump of the public acceptance line in future, as shown in Figure 14.4 at t_1.

If the company now consistently follows the principle of minimising its individual costs or risks, it will basically strive for solutions that will relieve it of its obligations at minimum cost and also from the risk of having to bear possible costs of additional later measures.

For example, if the company is faced with the alternatives of performing the decontamination either today, i.e. between t_0 and t_1, or at a later date, e.g. between t_1 and t_2, it will, in any case, select the present time, provided –

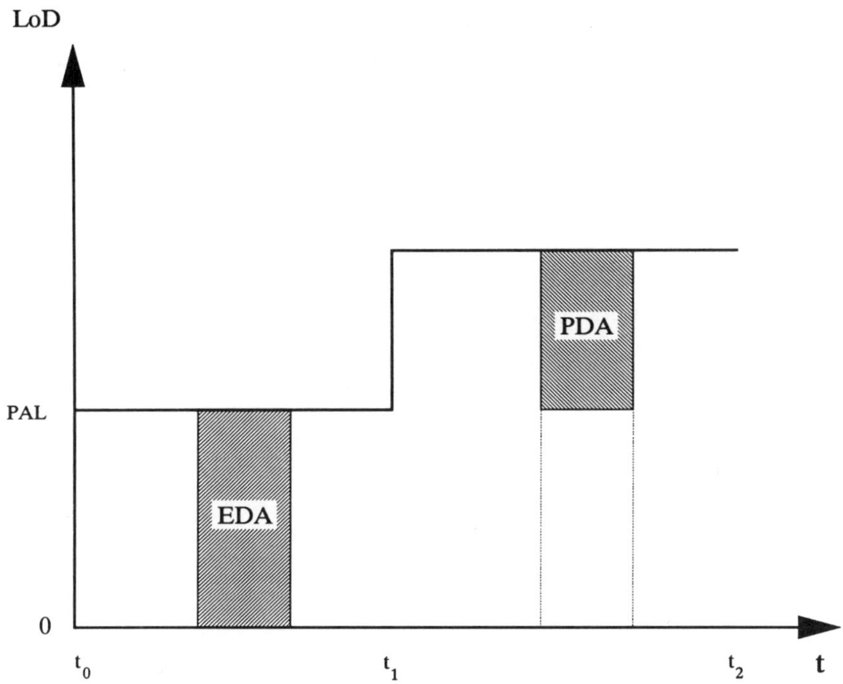

LoD : Level of Decontamination

EDA : Entrepreneur's Decontamination Activities

PDA : Public's Decontamination Activities

$\overline{t_0 t_1}$: PAL based on Conventional Technology

$\overline{t_1 t_2}$: PAL based on Innovative Technology

Figure 14.4 PAL-dependent cost distribution

and this is the decisive point – it is then relieved of all future obligations. The company will certainly accept the costs of the present measure EDA in Figure 14.4, and thereby avoid the extra costs of measure PDA which would be enforced under more stringent requirements later; these will have to be borne by the public. If, in contrast, the company waits until a time after t_1, it will have to bear both the costs of measure EDA and measure PDA.

More theoretically it may be stated that by its decision to carry out the measure today, the company transforms the relatively uncertain position of high overall costs (EDA and PDA) into a certain position of relatively low costs today (EDA). In other words, by internalising today's costs, the company externalises the risk of possible future costs at the same time.

Thus, this company utilises the respective (present and estimated future) levels of the public acceptance line as an element of its strategic behaviour by making use of information lags between the company and the public over time. Again, a public acceptance line based on technological specifications proves to be an environmental boomerang. For it does not lead to the intended complete internalising of the environmentally relevant costs by the creator (i.e. the company), but, on the contrary, to at least their partial transfer to the public. And if public funds are exhausted, the result necessarily is an insufficient reduction of environmental pollution.

The Significance of Environment-related Morality

At last it is necessary to talk about morality and responsibility. In case of purely individual economic rationality there would be in case of Figure 14.4 no reason whatsoever for the company to invest in more than just EDA. In the cases of Figures 14.1 and 14.3 there would also be only few reasons to select D or E instead of C (excluding, of course, the special aspects already mentioned above). Yet this insight is unsatisfactory. Moreover, this insight is shaken as soon as the category 'economic rationality' is broadened somewhat.

In our economic system, entrepreneurial activities are characterised by freedom of action. This freedom, however, is not unlimited. Freedom implies social responsibility and means that one is also accountable for the environmental consequences of one's own actions. And if one accepts the general applicability of the polluter pays principle (not least to set prevention incentives), the public will be permitted to insist on extensive internalising of external costs and risks by the companies even if prevailing laws and regulations do not enforce such internalising.

In the case of Figure 14.3, for example, public demand intends to achieve alternative D or E even though the given public acceptance line does not require it. And in case of Figure 14.4 this demand means voluntarily incurring both EDA and (at least part of) PDA costs in period t_0t_1. It should be emphasised that this is attributable solely to the entrepreneur's responsibility.

But normally the mere appeal for responsible action does not suffice to change entrepreneurial actions which are orientated purely on economic considerations, i.e. to minimise costs or risks. Instead, the reference point must be the economic interest of the companies.

There are always certain psychological components which are incorporated in public acceptance lines besides concrete scientific and technological knowledge. If a public impression arises, whether justified or not, that entrepreneurial actions and innovations take place without sufficiently considering the environment, psychologically motivated pressure to tighten the public acceptance line may occur.

This pressure resulting from public mistrust can only be removed if entrepreneurs visibly abstain from taking advantage of certain economic benefits as soon as significant environmental disadvantages could arise. Basically this is a plea for environmentally motivated abstention from short-term profit maximising. For if such an abstention is demonstrated in a credible manner, public confidence can be created to a significant degree. If a company or a whole industry should succeed in gaining such confidence, increased public acceptance of entrepreneurs' actions and innovations can result in the future, precisely from an environmental point of view.

If, as a result, the levels of specific public acceptance lines are not tightened so strongly in future, this will produce a trend to better profit prospects for the companies in the long run. And thus environment-related morality will – at least in this respect – no longer conflict with individual economic rationality.

Summary

This chapter examined the consequences of public environmental standards for entrepreneurial innovation activities. It focused specifically on the relation between technological entrepreneurial innovation and environmental risks, paying particular attention to influences on entrepreneurial decisions in the relations between internal and external risks in the course of time.

The reference point was the model of a so-called public acceptance line expressing advance public acceptance of certain pollution levels. Consequences of changes in this public acceptance line for entrepreneurial innovation behaviour and for the resulting entrepreneur's internal and external risks were analysed. Then consequences of publicly specified technology for the entrepreneurial innovation risk as well as for the environmental risk in general were discussed. Finally, emphasis was laid on the significance of environment-related morality from the perspective of both short-term and long-term individual economic rationality.

Bibliography

Endres, A., 1989, 'Allokationswirkungen des Haftungsrechts', *Jahrbuch für Sozialwissenschaft – Zeitschrift für Wirtschaftswissenschaften*, 40 (1), pp 115–29

Gethmann, C.F., Streffer, Ch., 1989, 'Standards für die Umwelt: Begrenzung des Risikos zwischen Nutzen und Schaden', *Essener Universitätsberichte*, 3 (1), pp 9–15

Karl, H., 1988, 'Ökonomie öffentlicher Risiken in Marktwirtschaften', *Wirtschaftswissenschaftliches Studium*, 16 (5), pp 217–23

Klemmer, P., 1990, 'Gesamtwirtschaftliche Effekte ökonomischer Instrumente im

Umweltschutz', in Wagner, G.R. (ed.), *Unternehmung und ökologische Umwelt*, Verlag Franz Vahlen, München, pp 262–82

Pearce, D., 1983, 'Are Environmental Problems a Challenge to Economic Science?', in O'Riordan, T., Turner, R.K. (eds), *An Annoted Reader in Environmental Planning and Management*, Pergamon Press, Oxford, pp 253–64

Pepper, D., 1989, *The Roots of Modern Environmentalism*, reprinted paperback edition, Routledge, London and New York

Schmidt, R.H., Wagner, G.R., 1985, 'Risk Distribution and Bonding Mechanisms in Industrial Marketing', *Journal of Business Research*, 13 (5), pp 421–33

Siebert, H., 1987, *Economics of the Environment – Theory and Policy*, second edition, Springer Verlag, Berlin.

Wagner, G.R., Fichtner, S., 1989, 'Kosten und Kostenrisiken der Altlastensanierung', *Zeitschrift für angewandte Umweltforschung*, 2 (1), pp 35–44

Wagner, G.R., 1990a, "Unternehmensethik" im Lichte der ökologischen Herausforderung', in Czap, H. (ed.), *Unternehmensstrategien im sozioökonomischen Wandel*, Duncker & Humblot, Berlin pp 295–316

Wagner, G.R., 1990b, 'Unternehmung und ökologische Umwelt – Konflikt oder Konsens?', in Wagner, G.R. (ed.), *Unternehmung und ökologische Umwelt*, Verlag Franz Vahlen, München, pp 1–28

CHAPTER 15

Towards a Vernacular Science of Environmental Change

Timothy O'Riordan

The risk literature has been dominated for many years by authors trying to distinguish between 'real' risk as judged by 'experts' and 'perceived' risk as felt by 'lay people' (see for example, Covello *et al.*, 1983). This is false distinction, in that it seeks to separate two ways of knowing that should be integrated. One of these ways is through the mechanisms of conventional, positivist, science which proceeds by the tried and tested methods of verification (or falsification), replication and observation via experiment. The other operates via a more mysterious process of experience, intuition and social communication. This is the 'traditional' science of peoples set in particular cultures of knowledge and beliefs (see Douglas and Wildavsky, 1984). Both are powerful and permanent. Their real strength lies in their union.

In his chapter for this volume, Professor Burke confesses surprise and some irritation regarding the apparently 'irrational' reaction by some members of the public to the decision, taken by the Advisory Committee on Novel Foods and Processes, to permit a very slightly altered yeast culture to be used in the public domain. The view of the Committee was that this culture was sufficiently close to its natural form, and sufficiently well studied, as not to constitute any conceivable danger to the public. Though the Committee was not party to the way the matter was finally handled, it appears that nobody thought to publicise the authorisation permitting the use of the genetically altered culture in the open, public domain.

The view of the critics was not so much to dispute the actual decision as to complain that the decision was taken in secret, without any advance notice, and therefore that there had been no opportunity for dialogue. The view of the Committee was that this particular decision was, in essence, a scientific 'non-event'. The view of the environmental groups was that any genetically altered material should be introduced only after the most extensive public debate, allowing for all legitimate fears to be aired. The fact that there was no such prior discussion created an opportunity for the critics to exploit the 'lack of trust' in the way the Committee worked. Top down expertise, no matter how great, gives the impression of scientific paternalism. This tends to alienate the public from the very processes of scientific discovery that are designed to serve the general good. It is this alienation which separates the two traditions of knowing. That is why, in the world of imperfect knowledge, we need a *vernacular* science. Vernacular science is used here to denote a special mixture of scientific analysis, communication, and participation that both beckons and unites.

A Micro Case Study of Failure to Incorporate Vernacular Science

The environmental idea lies in the synthesis of these two ways of interpreting the natural world. On the one hand, 'environment' is perceived as a set of processes capable of being examined and understood by objective empiricist science. In this way, for example, the yeast culture is largely a set of microbiological phenomena obeying laws as formulated by the conventional scientific method. On the other hand, 'nature' is perceived as being mysterious, sanctified and very precious. In this context, to molest the natural world is an act of moral criminality. So the addition of an 'unnatural' substance becomes a matter for ethical, or political, not merely 'scientific' debate. The 'vernacular science' of the title of this chapter, is really the emerging integration of these two ways of knowing and understanding. This is part of the essence of modern environmentalism.

In West Runton, in north Norfolk, the newly commercialised water company, Anglian Water Services plc, is proposing to collect the sewage from a number of small communities and pipe the untreated effluent into the North Sea via a long sea outfall. (I am grateful to Lawrence Bradby (1990) for providing me with background material on this study.) The idea was to avoid the cost of both primary and secondary biological treatment by aeration and bacteriological processes, remove only physical material by coarse screens, and let the 'natural' cleansing power of the North Sea do the work. In this way the commercial company would save some £25 million on capital costs and over £2 million per year on annual maintenance costs.

In adopting this approach the company believed it was acting on the best scientific advice. The view of the marine biologists, upon whom the company sought to legitimise this low-cost solution, was that pathogenic bacteria would remain, for the most part, off-shore, and, in any case, would not live long enough in the cold North Sea waters to give rise to any demonstrable 'risk' to the bathing public. In support of this contention, Anglian Water Services plc claimed that 'no medical authority has identified any of our bathing waters as a health hazard, and we are guided by their responsible and professional advice'.

It is worthy of note that in pursuing the long sea outfall, the company has the support of the British government, who accepts that this meets the European Community requirements of 'best available technology without entailing excessive costs' as outlined in various Community environmental protection directives. In addition the company stands to gain reasonable profit by embarking on a relatively low-cost decision, yet passing on the added costs to the consumer through a sewage treatment charge that is not perceived as being too inflationary. In this practice, therefore, the monetary objectives of the government are protected in part by the commercial instincts of a newly privatised water company.

The scientific basis for the Anglian Water Services decision has not gone unchallenged. Various environmental action groups, both on a national scale, and subsequently at a local level, have championed the cause of a clean North Sea and pollution-free beaches. These range from Keep Britain Tidy, which was mostly concerned with litter removal, through the Marine Conservation Society that is primarily interested in improved pollution control, to Greenpeace which is using its North Sea campaign to promote the cause of forcing industry to minimise its waste at source, not at the end of the pipe.

Locally, the Norfolk branch of Friends of the Earth became activated. Led by an energetic campaigner with no official scientific credentials, this group has captured many of the fears and the frustrations of the local population. The issue here is twofold. Many local people feel angry that Anglian Water Services' decision is being taken without proper consultation, and without any care for their fears over future littoral health. Added to this is anger and a growing sense that the company is scientifically misguided, and morally wrong in taking its proposed action. On the scientific side, the FoE group have collected arguments that pathogenic viruses from human waste can persist for much longer than previously thought (Rao and Melnick, 1984). By absorption on to marine sediment, viruses can remain infective for up to seventeen months. This line is being used by the local population as further justification for more effective control over the eventual decision, and better linkage between the 'analytical' science of bacteriological pathology and the 'experiential' science of morality towards the natural world. What is interesting is to see how persistent and politically effective this small band of campaigners is in both forging alliances with the local community, and maintaining a high political and media profile despite scientifically unproven accusations.

This case study will be pursued in the pages that follow. The message for the present is that a decision, taken within the umbrella of national policy, but at variance with a growing, yet not legalised, belief amongst North Sea states that the North Sea should no longer accept treatable wastes in an untreated form, cannot be made in isolation from the moral and experiential beliefs of the general public. The new environmentalism, in part, involves a politicisation of these beliefs into the mobilisation of interest groups, and into the international obligations of private companies towards collective environmental well-being. Vernacular science is not just participatory science: it is a science that transcends national boundaries and generations.

Innovation, Indeterminacy and Environmentalism

Innovation that deploys technology or new substances or processes tends to evoke concern nowadays along the lines that not all the responsible scientific evidence is in, so deployment should be delayed until more research is completed. Coupled with this has been the emergence in continental Europe of the precautionary principle, namely that it would be prudent not to assume that predictable science can ever be achieved, so it is better to err on the side of caution. The precautionary principle is enshrined in European Community law under the general argument of anticipate and prevent.

We can propose that there are three kinds of uncertainly in science. The first is a function of *inadequate data*. This is a trivial form, as it can be overcome by better monitoring and observation, though obviously that costs money. The second is due to *inadequacies of understanding* rather than knowledge. This can be addressed by a consensus attempt to cross disciplinary boundaries both within and between the natural and social sciences.

The third form of uncertainty is of direct relevance to the theme of innovation and change. This is where the phenomena being studied are not capable of being analysed by the conventional scientific method because they are *inherently indeterminate*. This means that the processes through which they work are not modellable without recourse to distorting simplification. This kind of uncertainty cannot be eliminated. It is found in the current debate over global warming, it is evident in the disputes over the role of tropical forest loss in contributing to local, regional and global climate change, and it is observable in the micro case study of the West Runton pathogens.

Figure 15.1 portrays this set of relationships based on a typology of three kinds of risk analysis proposed by Funtowicz and Ravetz (1990). On the horizontal axis are various degrees of decision outcomes, in terms of the effect on society and economy if decisions turn out to be wrong. On the vertical axis are the three levels of uncertainty alluded to earlier.

In the lower left arena, risks can be analysed and managed by conventional science and by expertise. Broadly speaking this is the arena of first order uncertainty. This is the general area of clinical science where the methodology of assessment is unquestioned. In the realm of genetic manipulation or in the early stages of pesticide formulation, this order of risk management remains firmly in the laboratory.

In the middle arena, the consequences of being wrong are greater, the wider environment is to some extent involved, and the level of uncertainty becomes a function of the experience and reputation of trained risk analysts. Here is the zone of second order uncertainty, where lack of understanding is compensated for by reference to informed judgement subject to peer appraisal. This arena covers relatively 'minor' or 'circumscribed' environmental

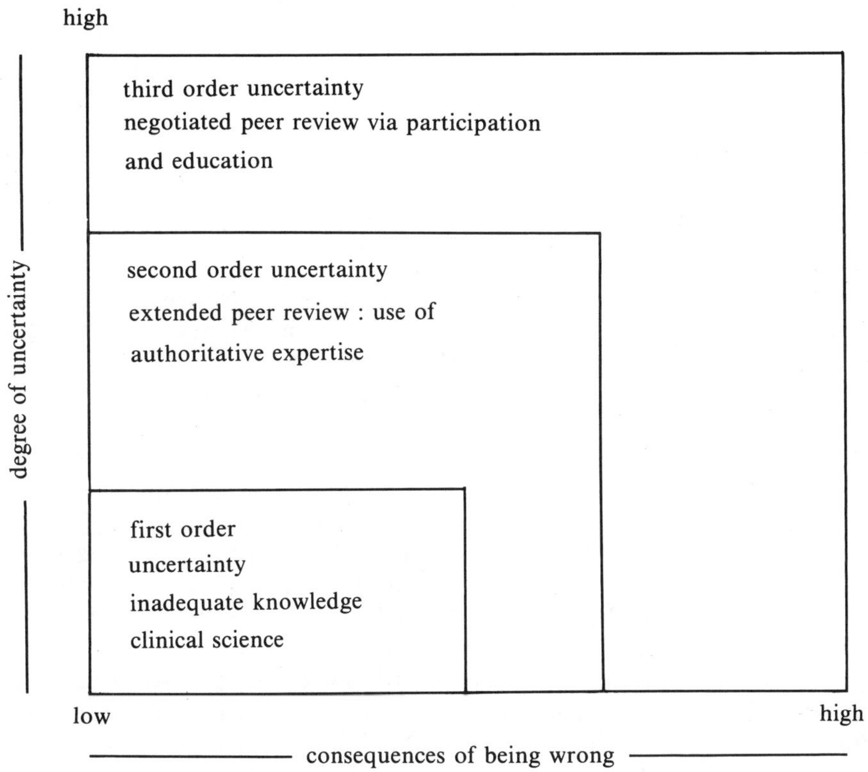

Figure 15.1 Levels of uncertainty

risks such as pesticides, and nitrates in drinking water. In some instances, although there is a groundswell of public concern, and though the precautionary principle may be invoked, in general the matter is dealt with by resource to 'authoritative expertise' of professional consultancy. The West Runton case falls into this category, at least in its early stages.

The third category of uncertainty is the one most likely to trigger the need for a vernacular science. This is the realm of systemic indeterminacy alluded to earlier. The need for vernacular science is most evident here, though it should be emphasised that the issues are beginning to enter the second-level category as well. Hence the emphasis on the West Runton case study to show how innovation of scientific 'awareness' is triggered by a particular form of decision pathway.

In this zone of systemic uncertainty, the mistake of being wrong could be serious, and the level of uncertainty is not only high, but not quickly narrowable. The issue here is partly a matter of methodology, and partly a matter of peer review. Here different aspects of science have to be

Table 15.1 Conditions for authority and styles of problem solving

Type of Uncertainty	Conditions for Decision Making	Prerequisite for Authority	Source of Authority	Problem Solving Style
Technical	Risk	Information	Instruction	Reductionist
Methodological	Uncertainty	Respect	Experience	Pragmatic
Epistemological	Indeterminacy	Faith	Revelation	Holistic

introduced. These involve ethical positions, consensus formation by public involvement and trust in participatory decision procedures. It is this last aspect which figures so prominently, when innovation, say in genetic manipulation, new chemical formulation, toxic waste disposal, or disputes over degrees of pathogenesis, creates a demand for a more 'user friendly' approach to science.

Table 15.1 outlines the varying mechanisms through which the three realms of scientific discourse are validated. In the arena of first order uncertainty the conditions for gaining legitimacy involve gathering more information by conventional empiricist means utilising low order concepts of risk management. The source of authority is hierarchical based on structures that disseminate information via instruction.

In the second order uncertainty mode, the prerequisites for legitimacy are respect for scientific competence, experience of offering judgement and pragmatism in attempting to solve problems. This mode of risk management is usually partially open to public intervention, though the dominance of respected authority is an important aspect of dispute resolution.

In the third order mode of uncertainty, actions operate on a mixture of conventional science and faith. The method of analysis is holistic, or at least interdisciplinary, and the source of final authority is revelation through participatory discussion based on shared trust between all concerned. In this instance, the precautionary principle is particularly important as a guide to action, and the task of the researcher is to try to design adaptive and flexible solutions that allow for adjustment as experience is gained and as knowledge increases. The demand for major investments in energy conservation as a safeguard against climate warming, even when the scientific evidence of warming is not assured, is an example of this approach. Energy conservation measures are a pragmatic and adaptive response, acting as an insurance against future uncertainty so as to reduce the premiums of a major salvage operation if the forecasts of climate-induced land-use changes and sea-level rises turn out to be right.

The use of the terms 'faith' and 'revelation' in this context can be misleading. The point is to show that problem-solving styles have to address both the context in which a problem is defined, as well as the manner in which acceptable solutions are found. The aim is to move the process of discourse from the 'wild' periphery, where antagonism and suspicion are rife, and where claim and counterclaim often result in unproductive dispute. The purpose is to focus on the 'tamer' centre, where there is hope of clarifying the underlying value positions and redefining factual uncertainties, providing a focus both for negotiation and for further research (see Rayner and O'Riordan, 1990). The interactive science involved in this process means seeking to unite the two sciences of 'environment' and 'nature', the interventionist aspect of human nature, and the respect for creation and the 'intrinsic' qualities of the natural world. The challenge for this extended form of science is to devise a mode of thinking and problem solution that extends the perspective of analysis from the domain of the individual as analysis here and now, to the individual as global custodian from now for ever more. This is the essence of the modern environmental debate : how to ensure that humanity can learn to sustain a world which it will always want to alter.

This may be an impossible goal, because it is difficult to conceive of any human-inspired course of action in which humankind is taken as the measure of all things. Those who want to dominate nature see in human action the scope for a world that is changed yet maintained. Those who want to revere nature pretend that it is possible to live in such a way that the maintenance element will be sustainable for ever more. The bridging mechanisms will not be easy to find. One avenue is for education via civic action, another is to create scenarios of outcomes which allow participation to become a negotiated mechanism for coping with endemic uncertainty.

Some Possible Common Methodologies

At least three groups of concepts can help bridge the gap between the natural and social sciences, and pave the way for vernacular science in general.

1 *Uncertainty, discontinuity and surprise.* In both the natural world and the social world, system states are not stable. Change may take place slowly by accretion on development or marginal adjustment. But change can also occur via topological flip, a sudden and unexpected disjuncture in a system state that results in a whole new perspective. Such has been the case with nuclear power since the early 1970s, and in radioactive waste disposal since the early 1980s. The aim is to build in improved capacity for adaptability in natural and social systems and to guard against avoidable vulnerability. Here is where participatory devices, including workshops, round tables, focus group discussions and scenario development may all play a role.

2 *Carrying capacity and evolutionary adjustment.* In the natural sciences there is a lively debate between equilibria processes and the scope for

organisms and communities to evolve in a 'negotiated manner' via opportunities offered by environmental change (see, for example, Goldsmith, 1989). In the social sciences, human organisations similarly evolve in relation to opportunity and stress. It may be possible fruitfully to interpret these concepts into a more reflective theory of organisational adaptation.

3 *Organisation, entropy, and in-built correction.* Following on from point 2 is the scope for using the principle of entropy, or channelling the internal energy of change. The key is the capacity for self-correction, which may be endemic, or may have to be structured from both within and from outside. It is this last point that should help to create the necessary capacity to cope with surprise.

The Dangers of Rushing

Vernacular science is rooted in an interdisciplinary science that not only merges conventional fields of knowledge, but also combines the ways of knowing in an age where 'science' still has many meanings. To get to vernacular science we must first build an interdisciplinary science based on the conventional modes of knowing.

Interdisciplinarity can only emerge from strong disciplines. Disciplines evolve as knowledge and understanding expand. But disciplines will endure, and must be allowed to do so. Interdisciplinary science will develop in suitable institutes and problem-solving frameworks, so long as those who are free and able to adopt the challenge are attracted in. This means good management which is suitably rewarded, peer review which is flexible but still inquisitorial, and the freedom to shift between disciplinary units and interdisciplinary complexes.

Even then, we must move slowly. There is no common language in interdisciplinarity. That will have to be developed by asking researchers to 'translate' concepts and principles with their own modes of analysis, then seeking common discourse. To rush into a common language too quickly could well prove counter-productive, distorting and alienating in unanticipated ways. The most appropriate solution is to experiment with care, build in self-correcting criticism, and operate through the common languages of the creative arts, as far as that is possible and appropriate. This collective examining could then become the bedrock of a more grounded interdisciplinary science.

To see how this interdisciplinary science can become extended into a vernacular science, we need to examine the changing character of environmentalism. Figure 15.2 illustrates its basic components and its progression of ideological positions. This diagram appears elsewhere so will not be elaborated here (O'Riordan, 1990). Suffice to say that the two ways of

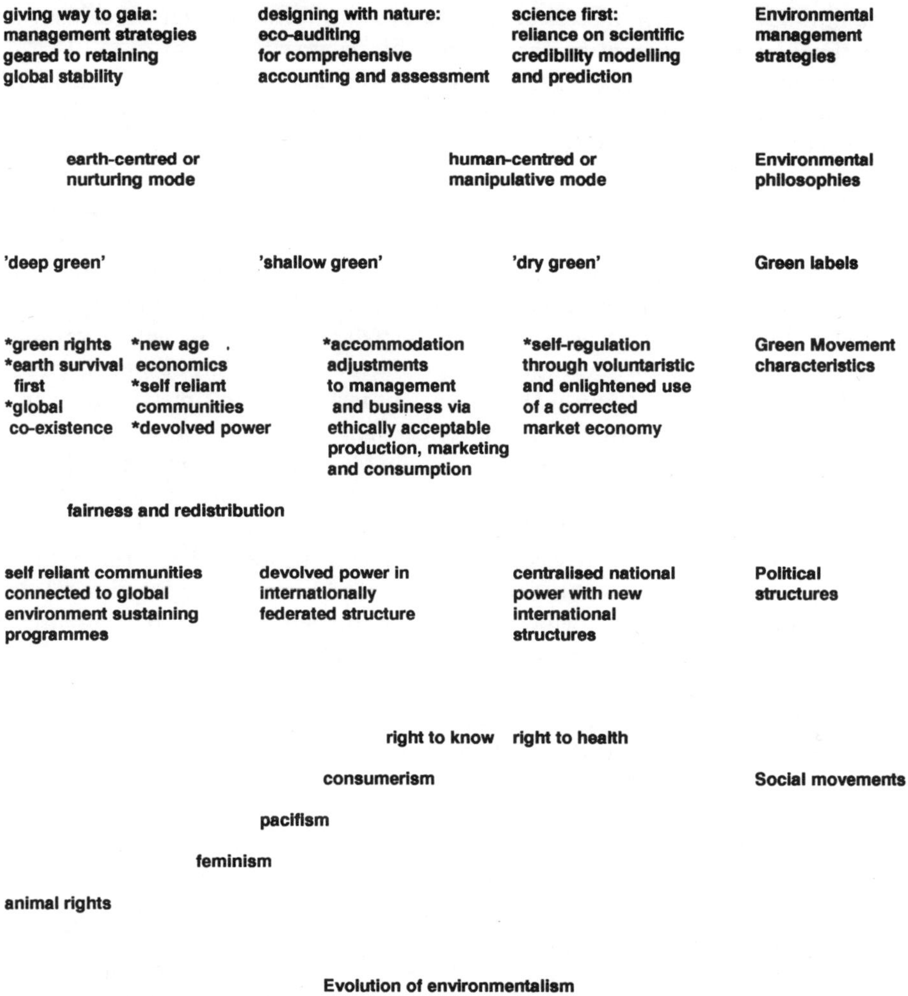

Figure 15.2 The contemporary dynamics of environmentalism, social change and environmental management

interpreting environment and nature result in three 'styles' of being 'green'. It is important to emphasise that each of these labels applies to a 'pure type' not to a given individual. Most if not all people reflect an amalgam of all three 'green identities'.

The dry greens see the world largely in instrumental terms. They believe that the processes of manipulation and innovation along lines that permit intervention in national systems can result in a better society, a transformed nature, and a self-educated citizenry. The enlightened self-awareness is encouraged by appropriate use of pricing and regulation and faith in adaptive management. Many of the examples involving genetic manipulation, radiation and pesticides reflect the dry green philosophy. Dry greenism is perfectly respectable: it is the essence of responsible innovation. But it confronts the two deeper layers of uncertainty and ethos outlined earlier.

The shallow greens are more adaptable and reflective. They envisage a form of intervention that is consonant with the natural world, allowing it to guide good practice. This is the core of the sustainable development philosophy: business broadly as usual, but with a 'green' face. Possible distortions or mistakes may be overcome by recourse to extended cost/benefit analysis, environmental assessment and risk management. Shallow greens, like the dry greens, seek no fundamental shift in the order of power and politics, though they conceive of a pluralist world in which varying reconstructions of power and pressure can form according to need.

Deep greens view a world that is ancient in origin and timeless in future. Human tenure is fleeting and ephemeral. The verities lie in the dramas of creation and evolution. Effective sustainable living can only emerge via sharing and redistribution of effective power to the community level. In risk terms, this means participation with a view to having effective influence over the control of health and well-being. Issues of justice towards the natural world, and fairness towards future generations become very significant. In this world, innovation which is not conducted via the participatory, negotiated means of a vernacular science, will be stoutly resisted.

Vernacular Science and the Mobilisation of Pressure

Figure 15.3 tries to show that what is described above is revealed in the work of the pressure groups in local, national and global arenas. The diagram indicates how the three 'green labels' become transformed into three kinds of environmental group. One is local or protectionist organisations, often known as 'NIMBY' groups. Ray Kemp, in this volume, outlines their changing role in his analysis of how the public respond to a well-intentioned consultative exercise over the siting of radioactive waste. NIMBYs start in a dry green mode, but, as in the case of both the NIREX experience and the West Runton study, they may move on to a more radical or transforming sphere where a greater sense of civic responsibility drives concern and action.

Shallow greenism tends to activate a second category of pressure. This emerges via civic-minded groups. They look for a better community in which to live, and seek access to decision making in order to ensure an element of power over how risks, or other disamenities, can be assuaged. Their aim is therefore to seek an element of trust and fairness in the process through which innovation is introduced.

This may mean quite demanding contracts with a risk-creating agency, such as a nuclear or toxic waste disposal organisation. It may mean the establishment of a community risk office staffed by independent yet

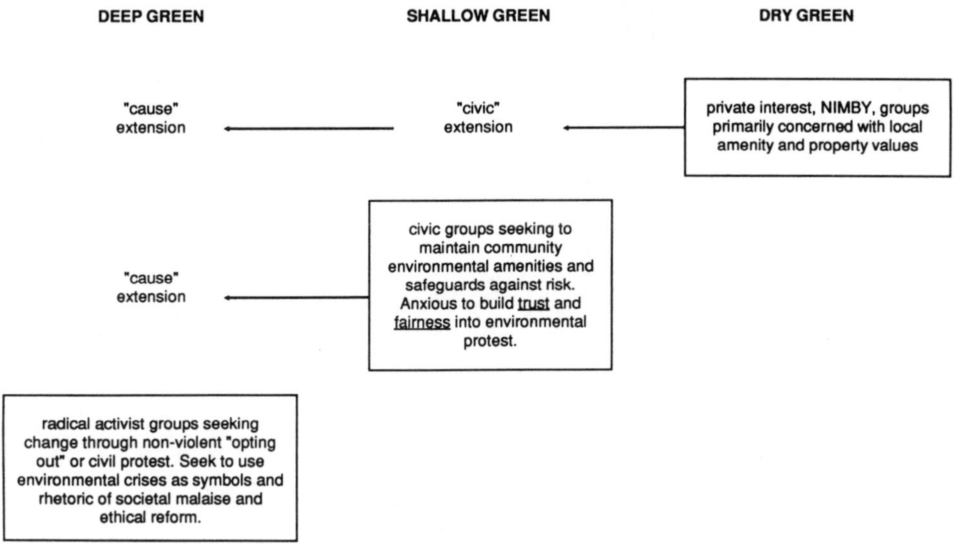

Figure 15.3 Typology of environmental groups

competent vernacular scientists, who act as a bridge between the source of possible danger, and the public anxiety. It may mean an extension of the absolute liability doctrine, whereby, through a communal 'fund' levied on the hazardous activity, compensation is available for all those affected in the event of an accident, payable on a 'no fault' basis subject to proper accounting.

The deep greens are represented by the more radical and activist 'cause' groups, some of whom seek a solution by 'opting out' of society altogether. Others use non-violence as a means of raising civil protest. They seek to use *causes célèbres*, such as a 'first of a kind' proposal, for instance, a nuclear station of a new variety, a toxic waste incinerator of a new technology, a waste water outfall which is deemed to create a 'new' hazard.

Figure 15.4 seeks to show how each of these groups, in entering disputes over innovation, attempts to adopt an education function by moving the passive, or unaffected, or unaware public into the politicised arena of protest and empowerment. This is done either by demanding new forms of control over risky technologies, or by requiring that essential moral stances about the effect of innovation or the natural order of things be taken fully into account.

Thus the civic groups, whether they are activated or NIMBY groups or simply extend their own participatory arena, seek to establish some sort of control over the nature and management of innovation. This may be achieved by tightening up regulations, by demanding some form of strict

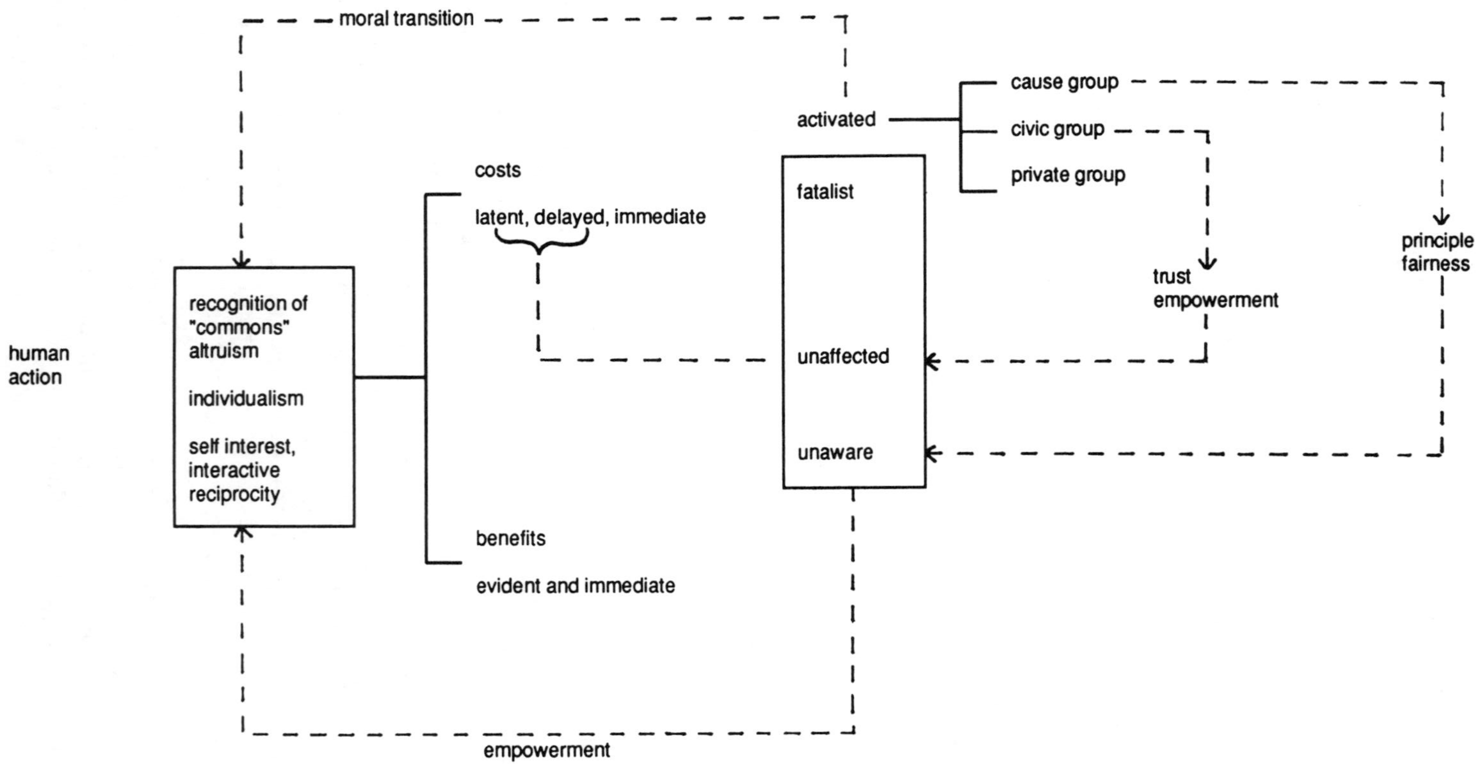

Figure 15.4 A typology of citizen action

liability, or by obtaining some form of compensation to recognise their 'loss of peace of mind' in accepting an alien technology or process in their midst. To achieve this they campaign on the second and third orders of uncertainty outlined in Figure 15.1 so as to activate the fatalists into believing that they can have power, and the unaffected into accepting a sense of communal obligation to the outcomes being negotiated.

This plan, or tactic, is evident in the disputes over the siting of radioactive waste facilities and the location of large toxic waste incinerators. It is unlikely that, in the absence of a national strategy for waste disposal, and in the absence of any adequate form of liability insurance to protect people on a 'no fault' basis in the event of an accident, that any compromise can be found over such disputes. In this context the second level of uncertainty needs to be supplemented by a communal form of education (in the broader sense of political awareness as well as learning) that permits citizens to become the basis of their own reference expertise.

The environmental cause groups seek grander outcomes. They genuinely try to get at the issue of morality, or principle, and so activate the unaware as well as the unaffected. Their aim is to change the basis of effective power, so that the more holistic and participatory modes of scientific discourse, particularly associated with the endemic uncertainty issues outlined in Figure 15.1, can be brought to bear on the very manner in which society participates in the process of innovation.

The West Runton story illustrates an amalgam of these two approaches. The role of FoE is to extend civic awareness into a moral crusade about how to treat the North Sea and devise appropriate norms of technological behaviour. They exploit the flux of conventional science, and the concern of many people, not just local, to want to care for the natural world in a more fundamental sense. At the same time they fuel their case with inflammatory support from a pliant news media which are prepared to promote the cause of sensationalisation and confrontation.

Imaging Futures

Disputes are not always constructive, but they can be educational. One way of combining the best of awareness raising in a non-threatening manner is to activate a more 'user-friendly' form of participatory science. This may be by way of a common language in art and literature as described earlier. One approach might be to model, say, future states of the North Sea coast on the basis of various scenarios of greenhouse gas warming, ill-combusted fossil fuel emissions, and forest removal. No particular picture would be accurate: quite the contrary, the range of scenarios would indicate the range of possibilities. By developing interactive computer programs, players could be invited to create their own futures on the basis of costed policy inputs. Thus they could see just what it might mean in prices and taxes to convert certain outcomes of global change into possible stable futures as sought by those who espouse the cause of sustainable development.

Bearing in mind the danger of rushing, it would be wise to treat such experimental formulations of imaged futures as a game rather than a matter of real policy choice. Pictorial images can convey rich amounts of information with small quantities of words. The interactive output then becomes the basis of learning and awareness raising. Hopefully, widespread experimental use of such techniques may result in an informed citizenry with a more active understanding of uncertainty and the scope for innovation in both technology and policy.

Towards a Name and an Ethos

The ideas above might, just might, result in a wider form of science. It is problematic to give it a name. 'User-friendly' is one term, 'vernacular' is another, 'negotiated' is a third. All suggest an interactive science of modelling and intuition, of analysis and ethos, of instruction and participation. Innovation is the essence of evolution: it cannot and should not be impeded by ignorance and fear. It can be accommodated if the method by which it is generated and handled is beckoning, receptive and supportive to the goals of a society that is to survive in a world always on the brink of turning hostile.

References

Bradby, L., 1990, *The politics of power: a West Runton case study*, School of Environmental Sciences, University of East Anglia, Norwich

Covello, V.T., Flamm, W.G., Rodericks, S.V., and Tardiff, R.G., (eds.), 1983, *The Analysis of Actual versus Perceived Risks*, Plenum Press, New York

Douglas, M., and Wildavsky, A., 1984, *Risk and Culture*, University of California Press, Berkeley

Funtowicz, S., and Ravetz, J., 1990, *Global Environmental Issues and Second Order Science*, Council for Science and Society, London

Goldsmith, E. (ed.), 1989, *Gaia and Evolution*, Wadebridge Ecological Centre, Wadebridge, Cornwall

O'Riordan, T., 1990, 'Major projects and the environment', *The Geographical Journal*, 156 (2), pp 141–9

Rao, V., and Melnick, J., 1984, *Environmental virology*, Van Nostrand Reinhold, Dordrecht

Rayner, S., and O'Riordan, T., 1990, *Chasing a Spectre : Risk Management for Global Environmental Change*, Council for Science and Society, London

CHAPTER 16

The Public Acceptance of Innovation

Final Session – Discussion Panel

Professor Roberts (Chair): May I say we are all very sorry that John Maddox isn't here to chair this session as we hoped he would be. This was always meant to be not only a discussion between members of this distinguished panel, but also a general discussion period in which I would urge everyone who still has questions to take part. At least two members of the panel have said that they would value the opportunity of steering the discussion by making some preliminary remarks. So, first of all, may I invite Professor Woolhouse to start off.

Professor Woolhouse: Just two points I think that I would like to make. One from the standpoint of my profession as a biologist, and the other from that of the level of an ordinary citizen commenting on social issues. The first thing I'd like to talk about is the question of predictability. I do want to underline that in this rising tide of work on genetic manipulation – and I don't think there's any harm in picking out this particular topic in risk assessment because it's only just at the very beginning – when you think of the millions of organisms and the enormous scope for manipulation, it's as though the piano had just been discovered and all the sonatas were there to be written as people might willy nilly wish to recombine things. I do want to emphasise that the complexity of the genetic make-up of an organism is such that we shall never get into a position of being able to predict totally the consequences of what we are doing, and the best that we shall be able to do is controlled release into limited areas. We can then, if necessary, sterilise and explore the propensity for invasion or behaviour within that limited context, and then, on that sort of evidence, assess the case for wider release. It's no use pretending that in the complexity of a biological system, that when you get down and interfere with the basic machinery, that you are going to be able to predict the consequences, any more than you can predict the overall spectrum of consequences in a much simpler device like a Cray computer when you put in a faulty chip.

The second point that I would like to comment on relates very strongly to something which was being said over dinner last night, and which was implicit in what Tim O'Riordan was talking about (Chapter 15), and this concerns the psychology of the public perception, because I find that this is something that has impinged on my own experience of life and which I am much puzzled by. I don't think it's that the white-coated scientists were appreciated in the 1950s but not now (as was said yesterday), but rather that the problems of technological society, and the associated worries which it generates, go back much earlier. As a little boy I used to be highly amused by Charlie Chaplin's films. I'm still greatly entertained by them, but for a number of different reasons which of course relate to their social comment on a technological society, and it does seem to me that this question of a feeling of alienation is something that is not one which can be resolved simply by tinkering with committees and representations on committees, but is something that is deeper in social concern.

I can tell you for example that in my own family, with my wife and daughter, the type of shopping basket described by Douglas Georgala (page 64) generates a feeling of abhorrence and suspicion, not only in the underlying technologies, some of which they are familiar with, but also because of a feeling of separation from what is natural, of the kind which Tim O'Riordan referred to. Obviously, I'm very much concerned with food production, and with sustaining massive populations, but we ask ourselves how much of that food manufacturing is really driven by public demand, rather than by mass persuasion through advertising, and how much is really intrinsic to a wholesome way of life, I think we have to face some serious problems. I think some of the alienation in the face of technology is because the forces that are driving technology are not those of simple human advantage – it is being foisted on people for other motives – but this is a very complex issue. It's one that I can't comment on as authoritatively as could many here, but in my perception it's a very important one indeed, and I would hope that it is one that the discussion might tackle.

Alan Mackley: I was once a technical manager in the commercial department of a major oil company, and I would like to take up this question of communication between companies and the public. I share the concern above, and I believe that one important channel of communication has been quite compromised by the form of information that is passed along it by the commercial management of companies to the public as consumers. I think you have to recognise that the same commercial management is determining, or strongly influencing, the allocation of resources within companies as well as influencing the way in which they communicate with the public. Reflecting back to the adversarial model (p.25) which crops up in society so much, whether in the political system, the legal system, or when we use the word public debate instead of public discussion (where the objective is often to seek the primacy of advocacy rather than achieve what the scientist would regard as a consensus), we can see similarities.

However, I don't think the situation is wholly pessimistic because there are other lines of communication that companies use. People who are consumers are also members of other groups – as shareholders, as employees, and as product users, rather than simply consumers – and I think these are groups which have overlapping interests, have interests in common with industry, and I think that these lines of communication are less compromised than the commercial one to consumers.

The conclusion I would reach for technologists, engineers and scientists within industry is that, firstly they have to convince their executive management that some of their commercial activities are counter-productive, counter-productive in that they risk compromising other lines of communication including the message on health and safety and environmental protection which they may wish to transmit. But, secondly, they should encourage management to use their contacts with employers, shareholders and product users, to communicate constructive and objective messages about health, safety and environmental concerns.

Dr. Boehmer-Christiansen: I just want to make a comment directly on this. I've recently come back from a conference in France addressing precisely this question of communication between industry and the public. After lengthy discussion and presentation of huge amounts of data the final conclusion was that industry will not be trusted by the public, it's too late! The commercial pressures and the vested interests are so great that any industry which wants to communicate on safety and risk is not going to be believed. The situation may not be identical in Britain, but the same underlying questions remain. Because, if you cannot find an institution which is trusted, then ordinary people, including me, and many consumers, who don't have the technical knowledge to make up their own minds on everything, have to find somebody whom they can trust. If, perhaps, the scientists working for Greenpeace, or Friends of the Earth, were on the advisory committees and part of the decision-making process, I would have more confidence that someone would befriend the environment rather than just economics, industry and science. Then, I might trust the outcome and judgements made by those committees more.

So two questions really: who do you think in British society is trusted at the moment so that people will believe what they say about risks, safety, and environmental and personal threats; and secondly, where does the opposition come from? Why should there not be laymen and laywomen, including environmentalists, on these advisory committees? They may not have all the scientific knowledge, but then nor do the scientists have all the social or economic skills. So what is the objection to having non-scientists on these advisory bodies on food, health and pollution?

Professor Roberts: Perhaps I may make one comment on your second question from my own experience. I have been in the past a member of committees that have tried to expand their membership to include just the constituency that you mentioned. The exercise will only be successful in my view if everyone concerned agrees that there is a consensus to work to. If there isn't that degree of common purpose, then it will fail. But I'd like other people to comment.

Professor Bobrow: This is getting very close to the only substantive point that I would wish to raise. I find it extremely interesting from a very narrow viewpoint of a medical and scientific background to think about the type of statement that is made as to why the public no longer trusts scientists, and the point that I came to when considering this was precisely the point that you have just made I think, but said slightly differently. It's not just that the public doesn't trust scientists any more. People also don't trust the church, teachers, politicians, the chap who fixes motorcars, Uncle Tom Cobley and all. So the real problem is not that we need to identify a group of people who are trusted, it's that we actually don't seem to have a mechanism for delegating trust in our current political and social organisations in a very effective way, and one possible cause for this, since I'm going to start on the presumption that human nature is no more cussed and difficult now than it was several hundred years ago, is simply the amount of information that now exists as compared to one or two generations back.

And I wonder whether a reasonably, well-educated young gentleman down from some university or other at the turn of the century might have felt that actually most of the important decisions taken in society were more or less within his comprehension. He may not have known the detail, but he could more or less have followed it. Whereas nowadays I think that there is no question that we all know that we are lucky if we can keep a grasp on one very small aspect of human knowledge, and therefore we actually need to be able to say quite formally, I do not understand anything about the real basis of that, and I'm going to have to delegate the factual basis of that knowledge to another crowd of people. And we need a mechanism for making that delegation, and I don't think we have it. It is clear that expert committees are not a sensible form of delegation in the general sense of the word. We delegate a lot of power through the standard political machinery to parliament, but politicians, although they have many uses, I won't say values, but uses, are again not a suitable repository of delegated trust on technical decisions. I think there's a vacuum in learning how to trust, rather than in trying to identify a group of people whom we do trust.

Lady Platt: I would agree with the last speaker. I don't think you can expect people to trust anybody. I think there is a healthy scepticism. Perhaps that's a product of a well-educated society, and also perhaps I think one of the best things that has been expressed in this conference is a sense of humility about not knowing everything, because nobody can. It has got to be teams of people, and that's why when you said is there anyone who is trusted, I was reminded of a Public Inquiry on planning. The fact that so many people are prepared to give evidence, and that the final report is the inspector's report, and is not a political matter, is crucial. So I think that that is possibly trusted, but otherwise I think that any decisions of this sort have got to be taken by a wide variety of people. I don't actually think you'll ever be lucky enough to have consensus, but then I think that part of the exercise of this healthy scepticism has got to be a dialogue, and everybody listening to other people and thinking 'well that's a point of view I haven't considered', and trying to take it into account in their own specialism. It's a synergy of specialisms.

Just recently the Human Embryology Bill has gone through the House of Lords and, as I recall, they said that the licensing authority should have not less than a third and not more than a half of its members drawn from lay people. Now I agree, I don't know how you are going to define lay – but what it meant was 'not expert' and 'not with an interest in it'.

Obviously I do have a special row to hoe because there are so few women on a lot of committees, and indeed there are some committees without any women at all, and I just don't think that's a possibility because they are ignoring the views of half the population, and however hard men try to put themselves into the minds of how a woman might view it, I don't think it can be done. This is something which experts from the food industry, given women's involvement in cooking, food buying and health, might like to follow up. And so perhaps that's just a contribution to the solution. Can I pass that on to others for comment?

Professor Georgala: I am pleased to say that women are now becoming more strongly represented on food advisory committees, even though this may be late in the day.

I would also like to say that in the UK there certainly are signs of significant trust in the food sector, not necessarily with manufacturers, but more on the retail side. There it has been possible to witness a high standard in terms of trust, good relations, quality and safety, and therefore I think it's not all hopeless. Retailers are, of course, in a very special position, but they have demonstrated a very high response rate to consumer worries, which is part of the nature of their business, and I think this may be an important feature of future commercial activity. In due course, it could feed back to manufacturing too.

Dr. John Sheail: Chairman, I wonder if I may make two points, I hope briefly. The first is that it seems to me that there is an anomaly in that, on the one hand, industry tells us that it finds it very difficult to get across the message that there is an industry that is concerned with the well-being of people and the well-being of the environment, and yet we know that that self same industry is mounting extremely effective advertising campaigns to sell its product, not only through the printed word but on TV. It seems to me there is a very complex issue here that industry and the community at large need to address in terms of the communications industry and the role of the advertising industry, a very complex issue relevant to our discussions today and yesterday.

The other point Chairman is this. I have the privilege of being Chairman of the governing body to St Peters Comprehensive School in Huntingdon, and I know if my colleagues on the teaching staff had been here yesterday they would have been extremely encouraged to hear reference to the role they are playing, and certainly should play much more, in terms of introducing children to some of the issues – the challenges we have been looking at over the last day or two, but I know my colleagues would say, 'Please, we need resources to do our jobs better'. For example, I'm conscious I ought to be in a finance committee this afternoon trying to think through how we can close our deficit without losing staff this coming year.

Clearly, local industry has a vital role to play in the relationship between school and industry, and this is one we are working on hard. But I was extremely discouraged the other day to be told by a governor whom we co-opted on to the governing body under statute, because we were trying to bring representation of the business community into our governing body with its enhanced powers, and that lady brought with her a keenness and skill we badly needed and the experience of a company of great stature in the community, that she felt she would have to resign because her company was not prepared to allow her to spend two or three afternoons a month in the school. It rather goes against the grain telling anecdotes in this country, but I find that very distressing indeed, and it does seem to me again a problem that needs to be addressed very urgently indeed. It's a problem that transcends the topic of our conference, but on the other hand, I also think that it has some relevance.

Lady Platt: That's what Industry Year was about, of course, and because of Industry Year there are a lot more school governors from industry. I would have said that it was worth while the Chairman of the governors going to see the Chairman of the local firm and saying this really is a very shortsighted point of view, and there is of course the right which might be quoted of any employee to have up to, I think, twenty-five days a year to take part in public life.

Dr. Sheail: Thank you for those comments. I should add that I understand that the case we are facing, and to which we should respond, is by no means an isolated one.

Professor Beringer: I am responding to your question of how to get people who are respected on to committees representing national interests and the public. This is something we believe is very important. As a result, we now have an environmental health officer who has been on the committee for just over a year as a non-expert, representing public interests in local environmental health issues. We also have a Nature Conservancy Council person, on the basis that the Nature Conservancy Council is meant to represent people who have an interest in the country looking nice as it is now. The difficulty is, as has been pointed out, that actually there are not strong, well-respected, national bodies that have really been shown to have their teeth into the issue of the preservation of the environment or whatever.

Professor Weale: I actually want to come in on another point. If you start looking at the membership lists of environmental organisations – take something like the Royal Society for the Protection of Birds – that's the largest wildlife protection organisation in Europe. Likewise, the National Trust has an extremely large membership. Now sometimes of course sheer numbers don't tell you much about the level of commitment. I think a lot of people within the National Trust probably fall into the same category that I do, which is that you are a member partly because you support the cause, but partly because it's a very good way of going to see country houses more cheaply than you otherwise could if you weren't a member. But I do think that there is quite a lot of evidence to show that there's a significant body of concern out there which is not adequately being tapped by our institutional resources.

Can I make one quite different point but which does relate to this, which is, I think, quite relevant to these discussions about the control of innovation and its relationship to public acceptability? These issues are usually phrased in terms of how are those who are concerned with the process of innovation going to make it acceptable to the public? In other words, because the source of innovation is so much a commercial and scientific activity, the question is always phrased as one about what sort of constraints does public acceptability put on the process of innovation. I think it's extremely difficult, but I also think it's very important, to try and devise institutions which might enable the public to signal the directions in which innovation might take place. It seems to me that there are clear directions in which the public would welcome innovation. One only has to think, for example, of the unexploited technological capacities in relationship to the physically handicapped or the mentally handicapped. These are directions of potential innovation which, if effort and resources were directed towards them, might actually be welcomed by the public. So I think it's important to bear in mind that there is necessarily perhaps an institutional asymmetry in the way in which we think about these things. Because the source of innovation doesn't often come from the public, the public is seen as a constraint on innovative capacity. If we actually thought more about what it was the public wanted, in terms of its values, and particularly vulnerable groups and so on, where there isn't currently much commercial or political pressure to get innovation in that area, it might begin to look rather different.

Professor Roberts: Thank you. May I invite you to say another sentence or two – since I am very sympathetic to the point you have made – about what sort of institution you think we should be looking at – even in principle?

Professor Weale: Well, I can see I've let myself into a situation where I've got to make some institutional innovation rather quickly! But let me just say, really very much off the top of my head, what an institution might look like. It wouldn't be impossible, for example, to think that within the public budget we could have a research and development fund which might be competed for by voluntary associations who are concerned with the welfare of vulnerable groups, in the sense that they could say, 'Well, we perceive a need here which can only be met by technological innovation – we would like to put up a proposal against this common fund that work be done in this area to see whether it's possible to develop aids for the visually handicapped or whatever it might be'. So I think it would be possible. One of the things that I'm struck by, doing the work on health services that I do, is that there is a tremendous amount of voluntary activity in the health care field which is concerned basically with the welfare of different patient groups, much on an under-resourced basis, which I think would be a potentially important source of ideas to see where we might want to innovate as a society and actually create new technologies which people would find valuable and which people would support.

Professor Allen: I have been trying to demonstrate the principle of humility and not say too much this afternoon, but I think basically the Interdisciplinary Committee has been starting off from a position within industry which leads me to a certain perspective. You see, the industries have an enormous momentum – they have a tremendous momentum – and the thing that I think we were trying to grapple with when we started to formulate this conference was to try and see how we could put in place the very enormous momentum of the industries. You can define the momentum of all our industries in relation to their product developments, the number of people working, the rate at which they are innovating – all this is clear – we know what it is. What we don't know very well is how rapidly these industries can change in relation to other feedback, such as from the environmentalists. And I think that poses the true question, and addresses the point that Professor Williams made which I felt was very significant.

He spoke about the need to have national purpose, insofar as we need to define those things on a large scale which can be identified, so that we can see how this mechanism is working. And if I can now suggest possibly how one might do it, it would be this. Take, for example, the power industry, oil, nuclear, gas and all the rest of it. In which way could there be a target for that industry which would be unarguably environmentally friendly and which could be a national objective? I would say the answer to this was wind power. You have to come down to some decisions like this, which specify new processes the industry could adapt to – and this very day in this University, there's the British Wind Energy group having their meeting. And we did a project actually on how this country could get 25 per cent of its electrical power from the wind. Now that seems to me to be a typically benign process. It's one which could be accommodated by our industry today, and it could be shown to happen. Now take another industry, transport. Again, a declared objective could be to have an environmentally friendly car. I'm already working with Sweden and the Shell Company and the Lotus Company in defining how to do that. And if one could devise some sort of tangible project or specification, that again would come into the list. Then, if you look into aircraft, there are definable new classes of aircraft in that transport system which could be declared and taken up, and so on. In terms of robotics – there is robotics work for the handicapped – there's a burgeoning enterprise in the robotics area, there are domestic robots, there are heroic robots for firefighting, and so on. So the point I want to make is that, apart from the general wish to have a properly constituted regulatory body, we have got to have some tangible targets.

Now another factor which somebody's got to respond to is that if there is a significant downturn in military business, big companies like British Aerospace, which includes Royal Ordnance now, will have to take quite hard decisions if they are going to avoid a significant diminution of orders. Where are they going to put their effort? Now, that could work in your favour, and bring these ideas closer to reality. I hope this has brought out Professor Williams' point, that we need some tangible, significant objectives, as an aid to how we somehow get this act together, and in response to the point that we do want to have some decisions or some activities that we can trust.

However, if you really want my view of why we are in dead trouble, I sometimes speak about signal to noise ratio. What I've just been describing are what I call signals. These are big signals of where we are going, but behind it all there's noise, and it's not so much noise *per se*, because you'll always get noise – it's whether the noise is destructively unstable that bothers me. Destructive instability of noise is very damaging, and I think perhaps this isn't scientific, but I think the fundamental troubles which have destabilised the confidence of people and prevented them from working out for themselves what they believe in, comes from the sheer complexity of today's life. It comes from the television, it comes from advertising, and probably also from *Which*? magazine.

Professor Georgala: Can I add, taking as an example the food industry, that we need to remember that lots of mistakes have been made. There have been industrial errors, poisonings, governments have permitted certain activities at one time and banned them ten years later. This has led to a genuine distrust of the system and I think we shouldn't assume that even if consumers were more fully informed, they would respond favourably. In fact, they may like the situation less and stop buying particular products – it depends what they learn.

I think we should realise there is a complex underlying situation in industry that is not all good. So industry perhaps has to change substantially, product ranges may have to change to go along with the evolution of communication.

Dr. Tait: We've been talking mainly about large multinational companies and how we can persuade them to adapt more quickly, but we haven't mentioned the role of the small company, and they are in fact very often much quicker on their feet, much more open to change. They can make decisions more rapidly and they are often more likely to develop environmentally friendly products for small markets than the big multinational companies. Yet we have the situation where most of the regulatory systems which are operating in the European Community are acting as a significant barrier to entry for small firms. Small firms can't afford the resources and the time commitments to meet the regulatory requirements, and I think this is a major problem that we need to think about when we design regulatory systems. Either we find a way to stop prohibiting these small companies, or we find a way to help the small companies over that hurdle so that they are not prevented from entering this market.

Professor Weale: On that point, I've been very impressed by some of the work that the Dutch regulatory agencies have been doing at provincial level, particularly on environmental protection, and there they are trying to develop schemes which entail going around and actually helping and advising small firms like garages and so on with their disposal problems, and positively helping them with the technical information which would enable them to improve their environmental performance.

I think one of the things actually we are terribly bad at doing in terms of innovation is of course innovating in terms of institutions. We find it very difficult to experiment with different institutional forms. There are all sorts of pressures to do with the nature of political systems and so on, which tend to push institutional reforms in a uniform direction, but I am very struck by our willingness on the one hand to regulate rather tightly innovative activity in the economic sphere, yet we seem to be quite happy with major institutional innovation without any experimental processes going into that at all, and it does seem to me there's a sort of mismatch. We don't think terribly creatively about institutions. Now maybe human beings aren't terribly inventive and there aren't very many different ways of regulating for environmental protection, or health and safety, but I'm not sure that all the ways in which we might do it have actually been thought of, and I'm very sure that not all the ways in which we currently do it have actually been thought through and experimented with.

Professor Bobrow: I get awfully confused on this question of communication. There are two kinds of communication, aren't there? There's the one where you actually go out explaining what it is you think you are doing and that's what most expert bodies actually do. They don't communicate in the two-way sense at all. They try to explain their thought processes in the hope that if only people out there would understand us, all would be well. Two-way communication implies that you have to be talking to someone else who is a representative of the public, and I keep hearing Friends of the Earth mentioned. But, of course, while FoE have a very valid point of view, and I have a lot of respect for them as an organisation and I have even, to tell you the truth, been a member, they actually aren't a major political party. There are major political parties. Why do you always talk about talking to Friends of the Earth when we should be talking to the major political parties, who, if anyone, actually represent the public of this country? If they don't represent the people of this country, then we have a really serious question to ask which goes far beyond all this.

To take this further, I think there are two kinds of conversation. It was epitomised for me by something which Roger Clarke brought up during his talk about radiation protection where he pointed out that there were a large number of different ways in which the International Commission for Radiological Protection can express health risks, all of them statistically valid. Now I thought that was fascinating. It carried the implication which I put to him and had confirmed that that same body of radiation protection experts were also going to now go ahead and choose, from amongst those large number of alternatives, which were the appropriate ones to use, and I thought that was a bad move. I think it is entirely appropriate for an expert committee to say, 'These are the ways which you could do it', but I think no expert committee ought to then make a social decision, because that's essentially a political point. And I think that's what the major political parties and indeed the pressure groups really ought to be distilling for us, and the fact that we do not always get clear feedback from the body politic as to what choices they want to make I suspect is because we don't ask the questions properly and we don't force them to make decisions. Politicians will not make decisions unless you force them. I think if we stand up and say, 'we'll take the decision for you', the implication being that we will also take the flak afterwards, they will say thanks and keep their heads down.

Professor Roberts: Dr Clarke has been telling me about an exercise that NRPB does to improve communication with some of the so-called pressure groups. I wonder if he would like to come in on this point.

Dr Clarke: We started an exercise about three years ago to try and have a dialogue with groups which had expressed environmental concerns about radiation. We seem to have successfully kept the meetings going for about three years. Initially I must say there was a certain amount of wary treading on the part of the representatives who came along, but we invited not only Friends of the Earth as everybody has mentioned and Greenpeace, but also professional institutions which had environmental interest branches, like the Royal Society of Chemistry, and, of course, people like the Consumers' Association, and I think we've had a useful dialogue. We've learnt from talking to them – this is talking to a restricted section of the community, as Martin Bobrow said, but they are a vociferous minority and the actual discussion has been very useful. Our Department of the Environment attempted to reproduce it elsewhere, an experience to which you, Lewis, were alluding, and it failed miserably, I think, because, as Martin said earlier, there was no identified consensus area to be worked towards. Nobody agreed or nobody believed that the Department of the Environment was serious about this; they thought they were just window dressing.

We would like to think that three years on we have achieved rather more of a rapport and a trust with these groups. I think we've learned from talking to them and probably to some extent modified our ideas, and I honestly think they've modified some of their ideas too. So the net result was that at one of these public inquiry forums, which Lady Platt mentioned, there was quite a useful and profitable discussion between the representatives of objector groups and ourselves, rather than a confrontation. We were both exploring tolerable risks and not violently disagreeing with each other, but searching from our different perspectives towards where an agreement might be found, and we hope that when the Inspector makes his report, just echoing what Lady Platt said, that that report will come out with some recommendations which we can all look to and accept as an honest, independent assessment of social acceptability, which we've all worked towards.

Professor Roberts: Thank you very much. But I must recall some of the history that I've been involved in where the reaction to public inquiry reports, excellent though they have been, has not been as friendly as that. The Inspector at Sizewell, despite his tremendous efforts, and the earlier Inspector at the Windscale Inquiry, which you will remember, ran into an enormous amount of flak from the environmental movement on the grounds that their concerns had not been taken seriously enough. And I think you are always up against this dilemma, whatever the rights or wrongs of a particular case: an Inspector has to make a recommendation. It will not, by definition, suit everyone.

Dr. Clarke: I agree with that, of course, but I was actually referring to the more recent Hinkley Inquiry. I think we had all learned from previous inquiries, and on both sides there were advances in the way cases were presented at Hinkley, and that's why I thought at the end of it, we might actually have made an advance in social judgement. I think positions were presented differently, on the basis of the experience gained at the previous, perhaps less satisfactory, inquiries.

Professor Roberts: I did not attend the Hinkley Inquiry myself. But I'm most encouraged to hear that. That's a very encouraging report indeed.

Dr. Boehmer-Christiansen: I should like to make just one comment on this, because I studied for quite a long time the decision making in industry in the 1970s. It really goes back to the earlier comment on noise. If you look at the state of the industrial process in this particular case, and I think it might apply to many other cases too, public opinion or public acceptability was really only tested or asked for at the very end, because nobody ever asked the public whether they wanted, in the early days of many of these industries, things like pesticides. This inevitably generated excessive concern from the environmental lobby as they were not consulted beforehand, thus concentrating and amplifying the anxiety. If we had a better industrial decision-making process – and we should be aware that our commercial institutions in many cases actually work in a most undemocratic environment, except the market, which is again right at the end of the chain when the product is finished – some of these problems could be avoided. Thus, much of the decision making is all left very much to the industrial sector. We have not really opened up the box in terms of accountability at all.

Professor Tivey: I'd like to make some general remarks about a certain feeling of unease over some of the discussion in which people seemed to be searching for someone they could trust. This seems to be a throwback to what was sought after in earlier years. I agree that the public seems not to trust the scientists so much now as they used to, but that's not too bad a thing. Why should they? They are still well ahead of the social services – and that's just about right, too. I did, to my amazement, hear somebody say 'trust the pressure groups', but no, they are no more reliable than anyone else. Then, to my further alarm, it was suggested we should trust the politicians to make our environmental decisions for us . . . no way!

This does seem to me to raise the suggestion that somehow there is a sort of almost Utopian vision arising which we are not going to have. Most of all I was rather disturbed by the person who wanted to determine objectives for society and this sort of thing, I don't like this idea at all. I would assume society is going to continue to be plural, and I would hope open, in which all issues of this kind would be continuously under challenge, and therefore we should not worry too much about the marginal lack of trust in scientists. That's beneficial to government. We shouldn't turn to somebody else instead, we should be continuously mistrustful.

Dr. Clarke: One of the things that I believe has contributed to this loss of trust in scientists, and many other groups, is the demand for speed of response these days. It touches on what Professor Williams said (Chapter 13) and what Professor O'Riordan said (Chapter 15), and the whole business of experts and advisory committees and the demands by politicians for instant answers to any new issue that arises. It's a certain fact that if I am not going to apprise the minister immediately on whether this is a good thing or a bad thing, you can be sure that somebody else will appear on television immediately saying whether it's a good thing or a bad thing, and some of the loss of public confidence I think is because of the conflicting views that are presented by 'experts', and I regret that I have no solution to this speed of response. I see it with the Stock Market. In the old days the Stock Market closed at 5 pm or so and everybody went home. They had a sherry in the evening, then they went to the club and they discussed what had happened in the day and they reflected. Now the computers are tuned into New York as soon as they go home. As soon as New York closes it's into Tokyo, and the computers are leading us through all the decision-making processes. We are swamped with information and we don't know how to handle it. And the net result is that you are expected to give an instant response when, as a scientist, you don't always know the exact answer immediately, and you need time to reflect, to discuss with colleagues, but unfortunately none of us are in a position where we are allowed that luxury these days. That has led to a lack of public confidence. It's led to a lack of correct decision making too, and as I said, unfortunately I am only too aware of the problem and don't see a solution. Perhaps I am wrong?

Professor Roberts: I'm afraid you've got a very important point there.

Professor Weale: I hesitate to say this because there are two very distinguished members of advisory committees to my right and other people in the room – but I do think it is worth trying to reflect on some of these issues, partly about trust, and partly about the speed of decision and the nature of uncertainty. It is of course true that it's of the essence of the scientific method that it be a process of conjecture and refutation, it's a process of actually allowing cognitive dissonance to take place, and not trying to suppress that, and at that level it seems to me that anything that's concerned with difficult decisions involving scientific assessments can't be based upon trust because the essence of the method is to challenge and to always be looking at assumptions.

I think there's a separate issue of trust which is perhaps related as much to the process by which we come to these decisions as to the actual outcomes, because clearly where there is controversy about what the outcome should be you can't satisfy everyone with respect to the outcome – that's in the nature of the case. However, I think what you could aspire to realistically would be to try and secure processes whereby people felt that the decision had been made in a fair and reasonable way, and in which their viewpoint had somehow been weighed. I suspect that one of the most important requirements is that people feel that what they say does genuinely have a chance of making a difference to decisions. And I think that the issues we are talking about are not straightforward scientific questions, nor straightforward questions of value. We are typically talking about issues which are highly complex technically, and in which many of the crucial decisions concern which test statistic you use as the basis for regulation, and these are not straightforwardly scientific issues at all. So I think that where the trust is going to be crucial is not in terms of believing what people say, because after all we want a process of challenge and open discussion. I think that where the trust has to come is in devising procedures under which the disagreements that take place are seen to be good faith disagreements, and not motivated by commercial considerations or political interests or whatever, but are genuine expressions of differences of view about what's in the public interest. That I think is very difficult to get.

The point I think I would make is that I think you are more likely to get a sense that a process has been a good faith process if it is not made under too much heat of controversy, and if I were looking to experiment with institutional forms here, I think that one of the things that I would say is, 'Well, let's try and get a process that generates some consensus of understanding in a difficult area when it's not the focus of public attention'. Because if it's a focus of public attention then the stakes become high very quickly and if it matters what the outcome is it's terribly difficult to make decisions. We might actually use this fact that the cycle of attention is very variable and doesn't always shine in the right places at the right time. We might say to ourselves, 'Well, let's at least try and get some process which generates understanding and consensus about that' because when it does become hot at least we can pull this out and say, 'Well, look, we have actually thought about this, and we've got some understanding, if not a complete understanding'.

Professor Bobrow: We should also recall the fact that almost all of this communication on the broadest level in fact takes place through the not-entirely-true-transmission of the public media, and it's clearly too late to discuss that, but it's as well to bear it in mind.

Professor Roberts: Perhaps you'll allow me one anecdote. I was trained for a TV appearance, and the guru who was training me, after about half an hour of constant effort, said 'Roberts, you are absolutely hopeless, you're trying to answer the questions; you've got to realise that this medium is a medium for the half truth and you've got thirty seconds to say it'.

We've had a good session this afternoon. We've exposed most of the important points that have come up in the last day and a half again, and seen our way through to some hopeful solutions coming from more than one speaker. It's not going to be easy, I'm sure; no one has thought of a magic wand and there won't be one, but at least I don't think we need end in a spirit of complete gloom. To close the Seminar I'd like to ask the Chairman of the Organising Committee, Professor Allen, to say the last few words.

Professor Allen: Thank you very much indeed for giving me the last say. I won't take advantage of it. I must say that we've learnt so much these two days and we've made so many good friends. I've been impressed with the real professional standards of the speakers. We felt we had done a good job in getting the best possible speakers we could and I feel we can be satisfied with that. But the audience is entirely a matter of luck, and we've been lucky because you can't persuade people to come – you can only send out a thousand mail shots – and the forty or so that turned up must obviously have wanted to get here and they have made it a really splendid occasion.

One thing in particular which I've learnt, is that when we talk about innovations, I suspect we are usually thinking of projects and products, but now we've heard from Albert Weale that we really need to carry over this concept of innovation into the national decision-making process itself. Like any innovation, there's nothing that's been like it before, and if we are looking for new innovative frameworks which will help the country get its industries reshaped, both big and small, and take much better cognisance of these various changing views, then the search for alternatives is going to be on our agenda and my experience of innovation is that at such a time of great change perhaps we will have to go to someone who is not an expert at all, someone who is quite outside the field, who comes up with a very cogent way of putting it into a format. If it's very innovative we may have to try thirty or forty different decision process mechanisms before we get the one that's right.

Also, I felt that although we only had one media person present, he did give us extremely good value, he was extremely honest, and I liked his idea of having a book together with a TV programme. I thought that was sufficiently novel to hope that it succeeds in getting us out of some of our problems.

In concluding this I would like to thank all the members here of the UEA who have been such excellent hosts. I don't think we have been short of anything including sleep, and I must really thank particularly Alison Clarke who has done all the hard work behind the scenes, in very difficult circumstances. But in the end it all came together on the day, on the two days, and I hope you have learnt something from it as we have, and if any of you wish to get the Interdisciplinary Committee to pull any chestnuts out of the fire for you, please let us know.

Appendix: Seminar Details

Keynote Address	The Public Acceptance of Risk and Innovation Baroness Platt, CBE, F.Eng.
Session 1	Chairman: Prof. J.E. Allen, The Interdisciplinary Committee Innovation and the Public Perception of Risk J.D. Rimington, CB The Role of the Media G. Lean Acceptance of Innovation: An Industry View of Environmental Aspects J.C. Felton
Session 2	Chairman: Dr. J. Bowman, CBE, Chief Executive, National Rivers Authority The Regulation of Pesticide Use: An Historical Perspective Dr. J. Sheail Pesticide Innovation and Public Acceptability: the Role of Regulation E.J. Tait, S. Brown and S. Carr
Session 3	Chairman: Prof. H.W. Woolhouse, Agricultural and Food Research Council The Release of Genetically Modified Organisms Prof. J.E. Beringer Health Risks from Food Prof. D.L. Georgala, CBE Evening: Public Acceptance of Innovation Prof. D.C. Burke
Session 4	Chairman: Prof. M. Bobrow, Paediatric Research Unit, Guy's and St Thomas' Hospitals

	Radiation Protection Standards: How are They Set and are They Adequate? Dr. R.H. Clarke Institutional Innovation to Generate the Public Acceptance of Radioactive Waste Disposal Dr. R. V. Kemp Variation in Information Use in the Evaluation of the Social Acceptability of Nuclear Waste Management Options Dr. P.T. Allen
Session 5	Chairman: Prof. A.P. Weale, Centre for Public Choice Studies, University of East Anglia. Innovation and the Political Context of Technical Advice Prof. R. Williams Entrepreneurship and Innovation from an Environmental Risk Perspective Prof. G.R. Wagner Towards a Vernacular Science of Environmental Change Prof. T. O'Riordan
Panel	Chairman: Prof. L.E.J. Roberts, FRS, Environmental Risk Assessment Unit, University of East Anglia. Rapporteur: Dr. D.J. Ball Members: Baroness Platt, Prof. J. Allen, Prof. M. Bobrow, Prof. A. Weale and Prof. H. Woolhouse

The Seminar on 'The Public Acceptance of Innovation' was arranged by the Interdisciplinary Committee in conjunction with the Centre for Public Choice Studies and the Environmental Risk Assessment Unit of the University of East Anglia. The Seminar was held at the University of East Anglia on 27–28 March 1990.

Organising Committee

Professor J.E. Allen
Cranfield Institute of Technology
Chairman, The Interdisciplinary Committee

Dr J.C. Bowman, CBE
Chief Executive
National Rivers Authority

D.R. Cope
UK CEED

Dick Crampton
The Interdisciplinary Committee

Professor T.R. Lee
Psychology Laboratory
University of St Andrews

Professor A. Porteous
Faculty of Technology
Open University

Professor L.E.J. Roberts, CBE FRS
Environmental Risk Assessment Unit
University of East Anglia

Professor A.P. Weale
School of Economic and Social Studies
University of East Anglia

The Interdisciplinary Committee had its origins in the early 1970s as an initiative of the chartered engineering institutes. Its remit is to survey the domain covered by forty-two scientific and engineering professional bodies for novelties that may not otherwise be attended to by existing organisations. The Committee consists of individual professionals chosen for their understanding of innovation and interdisciplinary issues. It operates as a catalyst for conferences and publications, and has tackled such subjects as the transportation of hazardous substances, man-made islands, economic recovery of deserts, lithium, technology in the Health Service, and this Conference at the University of East Anglia.

The Chairman is Professor John Allen and the Secretary is Mr. S.J. Crampton (10 Church Street, Sutton Courtenay, Abingdon, Oxon OX14 4NJ, Tel: 0235 848243).

The Environmental Risk Assessment Unit was established in the School of Environmental Sciences in 1986 to act as a focus for research on environmental risks which could become significant questions of public policy. The research programme aims to exploit the multidisciplinary strength of the School and to span the technical questions arising in risk assessment and the social questions involved in risk communication and risk regulation.

Address: Environmental Risk Assessment Unit, School of Environmental Sciences, University of East Anglia, Norwich NR4 7TJ, Tel: 0603 56161, ext. 2547 and 2838.

The Centre for Public Choice Studies was established in 1987 to act as a focal point for research among social scientists at the University of East Anglia on questions of public policy, public choice and decision making. Its associated faculty has undertaken research on such topics as integrated pollution control in Britain and West Germany, AIDS policy in the UK, the ethics of resource allocation in the National Health Service, and public attitudes to local authority services.

Address: Centre for Public Choice Studies, University of East Anglia, Norwich NR4 7TJ, Tel: 0603 592271.

Index